LIBRARY COPY

NUTRITION AND THE CLIMATIC ENVIRONMENT

STUDIES IN THE AGRICULTURAL
AND FOOD SCIENCES

Nutrition and the Climatic Environment

WILLIAM HARESIGN, PhD
HENRY SWAN, PhD
and
DYFED LEWIS, DSc

Faculty of Agricultural Sciences
University of Nottingham

BUTTERWORTHS
LONDON - BOSTON
Sydney - Wellington - Durban - Toronto

The Butterworth Group

United Kingdom London	**Butterworth & Co (Publishers) Ltd** 88 Kingsway, WC2B 6AB
Australia Sydney	**Butterworths Pty Ltd** 586 Pacific Highway, Chatswood, NSW 2067 Also at Melbourne, Brisbane, Adelaide and Perth
South Africa Durban	**Butterworth & Co (South Africa) (Pty) Ltd** 152–154 Gale Street
New Zealand Wellington	**Butterworths of New Zealand Ltd** 26–28 Waring Taylor Street, 1
Canada Toronto	**Butterworth & Co (Canada) Ltd** 2265 Midland Avenue Scarborough, Ontario, M1P 4S1
USA Boston	**Butterworth (Publishers) Inc** 19 Cummings Park, Woburn, Mass. 01801

All rights reserved. No part of this publication may be reproduced or transmitted in any form or by any means, including photocopying and recording, without the written permission of the copyright holder, application for which should be addressed to the publisher. Such written permission must also be obtained before any part of this publication is stored in a retrieval system of any nature.

This book is sold subject to the Standard Conditions of Sale of Net Books and may not be re-sold in the UK below the net price given by the Publishers in their current price list.

First published 1977
ISBN 0 408 70819 0

© The several contributors named in the list of contents, 1977

Library of Congress Cataloging in Publication Data

Nutrition Conference for Feed Manufacturers, 10th,
    University of Nottingham, 1976.
    Nutrition and the climatic environment.

    (Studies in the agricultural and food sciences)
    Bibliography: p.
    Includes index.
    1. Animal nutrition—Congresses. 2. Livestock housing—Congresses. I. Haresign, William.
II. Swan, Henry. III. Lewis, Dyfed.
SF95.N87   1976   636.08'52   76–45666
ISBN 0-408-70819-0

Typeset & produced by
Scribe Design, Chatham, Kent

Printed in Great Britain by Billing & Sons Ltd.,
Guildford & London

PREFACE

The tenth in the series of Nutrition Conferences for Feed Manufacturers focussed attention on the effect that climatic variables have on the efficiency of production of farm livestock. Whilst the general effects of these interactions have been studied extensively it is only very recently that the knowledge attained has been applied to practical production systems. The importance of this subject lies in the fact that artificial manipulation of the climatic environment, especially in pigs and poultry, can lead to greater efficiency of production by altering the overall energy balance of the animal.

The discussion commenced with a general paper on 'Environmental Factors and their Influence on the Nutrition of Farm Livestock' with special emphasis on the ruminant.

The theoretical aspects of nutrition—environment interaction in both poultry and the growing pig were discussed in detail, followed by papers on the more practical aspects of how to manipulate the climatic environment of these two species to improve the efficiency of production.

The meeting was closed by an important group of papers on 'The Nutrition of Rabbits', 'Calcium Requirements in Relation to Milk Fever', 'Protein Quantity and Quality for the UK Dairy Cow' and 'Nutritional Syndromes of Poultry in Relation to Wheat-Based Diets'.

All of the papers are written in a clear and informative manner and are likely to be of immense interest to all those working in the field of the nutrition of farm livestock.

Nottingham 1977

W. Haresign
H. Swan
D. Lewis

CONTENTS

1. **ENVIRONMENTAL FACTORS AND THEIR INFLUENCE ON THE NUTRITION OF FARM LIVESTOCK** 1
   K.L. Blaxter, *Rowett Research Institute, Bucksburn, Aberdeen*

2. **NUTRITION–ENVIRONMENT INTERACTIONS IN POULTRY** 17
   A.H. Sykes, *Wye College, University of London*

3. **CLIMATIC ENVIRONMENT AND POULTRY FEEDING IN PRACTICE** 31
   G.C. Emmans, *East of Scotland College of Agriculture, Edinburgh*
   D.R. Charles, *ADAS, East Midland Region, Derby*

4. **THE INFLUENCE OF CLIMATIC VARIABLES ON ENERGY METABOLISM AND ASSOCIATED ASPECTS OF PRODUCTIVITY IN THE PIG** 51
   C.W. Holmes, *Massey University, New Zealand*
   W.H. Close, *ARC Institute of Animal Physiology, Cambridge*

5. **CLIMATIC ENVIRONMENT AND PRACTICAL NUTRITION OF THE GROWING PIG** 75
   D.G. Filmer, *Dalgety Crosfields Ltd., Bristol*
   M.K. Curran, *Wye College, University of London*

6. **THE NUTRITION OF RABBITS** 93
   J. Portsmouth, *RHM Agriculture Ltd., Berkshire*

7. **CALCIUM REQUIREMENTS IN RELATION TO MILK FEVER** 113
   D.W. Pickard, *Department of Animal Physiology and Nutrition, University of Leeds*

8. **PROTEIN QUANTITY AND QUALITY FOR THE UK DAIRY COW** 123
   W.H. Broster and J.D. Oldham, *National Institute for Research in Dairying, Reading*

9. **NUTRITIONAL SYNDROMES OF POULTRY IN RELATION TO WHEAT-BASED DIETS** 155
   G.G. Payne, *Poultry Husbandry Research Foundation, University of Sydney, NSW, Australia*

10 FUTURE DEVELOPMENTS IN FEED COMPOUNDING   175
    IN EUROPE
    C. Brenninkmeyer, *Chairman of Committee A of FEFAC*

LIST OF PARTICIPANTS   183

INDEX   197

1

ENVIRONMENTAL FACTORS AND THEIR INFLUENCE ON THE NUTRITION OF FARM LIVESTOCK

K.L. BLAXTER
Rowett Research Institute, Bucksburn, Aberdeen

The effects of environmental factors on animal production are both indirect and direct; indirect in that they affect the amount of food available, direct in that these factors affect the energy needs of the animal. There is, however, an interaction between these two effects; that is, nutritional status affects the animal's response to its environment.

Though the United Kingdom is in the temperate zone, it is not immune to the effects of climatic variability. The UK may not experience the calamities of total crop failure as a result of adverse weather, but it has to be remembered that widespread crop failures due to climatic factors in overseas countries affect world markets for grains, oil seeds and protein concentrates. As a result price rises take place affecting British animal production. The poor harvest in the USSR is not entirely due to bureaucratic incompetency; partly it is due to poor growing weather. Whatever the reason, the USSR has now entered world markets for cereal grains and this will in part determine the level at which animal production here is pitched.

Even so, the environment here certainly affects animal production and two examples from the last year illustrate the indirect and direct effects. The hay and silage crops in the summer of 1975 were light and the drought conditions, particularly in England, certainly reduced pasture productivity and yields of crops, particularly of root crops. Had it not been for a fairly open autumn in the year there would have been serious shortage problems for winter fodder. The major factors in the United Kingdom accounting for the indirect effect of environment on animal production are undoubtedly all those that can contribute to a soil water deficit. In addition, too much water is equally important. Rain in excess in autumn reduces the acreage sown to winter corn, wet warm weather encourages fungal disease in crops such as potato blight, while rain and wind at harvest leads to laid hay and cereal crops, losses of quality in forage and in grain alike and with the latter usually a loss in yield.

These indirect effects might be elaborated, but it is more useful to examine the direct effects of climate. In this respect, despite the fact that summer 1975 was the warmest encountered for many years, there is no evidence that, provided drinking water supplies were adequate, livestock suffered any diminution of productivity as a result of the direct environmental effect.

2   *Environmental factors – influence on nutrition of farm livestock*

Turning to the second example, on April 8, 1975 a sudden storm occurred in Scotland. It was somewhat local; no snowfall of any magnitude occurred in Inverness yet in Banff and West Aberdeen bitter cold and snow created blizzard conditions for a protracted period. A man died on the Cabrach that night when he strayed from the tractor he was using to cart food to ewe hogs and was lost. Another farmer, farming land above 1000 feet, had expected a lambing shepherd to arrive but he could not get through the drifts. The flock was lambing and the farmer and his wife undertook the work in those appalling conditions. Losses of lambs and of ewes were so heavy, that this farmer compared the conditions to those in 1947 and 1938. His was but one farm and no-one can say what total losses were. The local knackery provides some clue since they stopped accepting dead ewes for the ensuing two weeks because they could not deal with the influx.

This second example of the effect of the environment illustrates the effects of calamitous weathers, short-lived and happily infrequent. It would be nice to be able to predict the probability of their occurrence as well as forecast their time of appearance. The same applies to longer spells of cold weather. The last few winters have been mild as judged by decades of experience; it would be useful to know what the next few months will bring so as to judge better some of the weather risks that are necessarily taken. Furthermore, much longer expectations of weather are needed so that decisions on investment in house design and insulation for in-wintered or intensively husbanded stock can be made. Estimates of the probability of occurrence of combinations of environmental factors and their duration should be given as much attention as short-term forecasting of the weather if the best use of information about the effects of these factors on animal production is to be made.

This information is now considerable for much has been accomplished in the last 20 years to define and quantify the effects of the physical environment on stock. An Easter School at Nottingham dealt with these problems recently (Monteith and Mount, 1974) and is an excellent source of reference. This chapter will explain the general framework of ideas which has grown up and within which most of the discussion will take place in subsequent chapters. Since cattle and sheep are not being accorded a separate section most, but not all, of the illustrations here will be taken from work with them; the principles are the same, however, whatever the species.

Units of Measure

Some difficulty may be occasioned in reading the literature of this subject because of differences between authors in the units of measure they use. Thus the heat production of an animal per unit of time has been expressed as calories, joules or even British thermal units per minute, hour or day. A joule per second is a watt and so heat production per day can, and indeed at the Easter School, was expressed as watts. In addition, rate of heat production has been expressed in absolute terms, per kg of weight, per kg weight raised to a power or per unit surface.

It seems sensible to agree on a common measure at least among those concerned with nutrition. The unit of time in nutrition is the day, and the joule is the unit of energy used. These are both now firmly established. Since heat exchanges of animals take place at their surface the square metre as the unit of surface area must be incorporated. By keeping to days, square metres and joules it is possible to move easily from tables of nutrient needs expressed in joules per day ($J\ d^{-1}$) such as that of MAFF (1975) to estimates of heat losses occasioned by the environment and back again to estimate the food costs of cold conditions. This approach will be illustrated later. Measurements of insulations in terms of $°C\ m^2\ W^{-1}$, or worse (see Monteith (1973) for a derivation), $m\ s^{-1}$, must be replaced by the more direct and more easily manipulated statement $°C\ m^2\ d\ MJ^{-1}$. The Appendix provides conversion factors.

Heat Production and Temperature

Figure 1.1 shows the effect of air temperature on the metabolism of a sheep closely clipped to maintain constant its fleece length (Graham *et al.*, 1959). Different amounts of feed caused different heat productions,

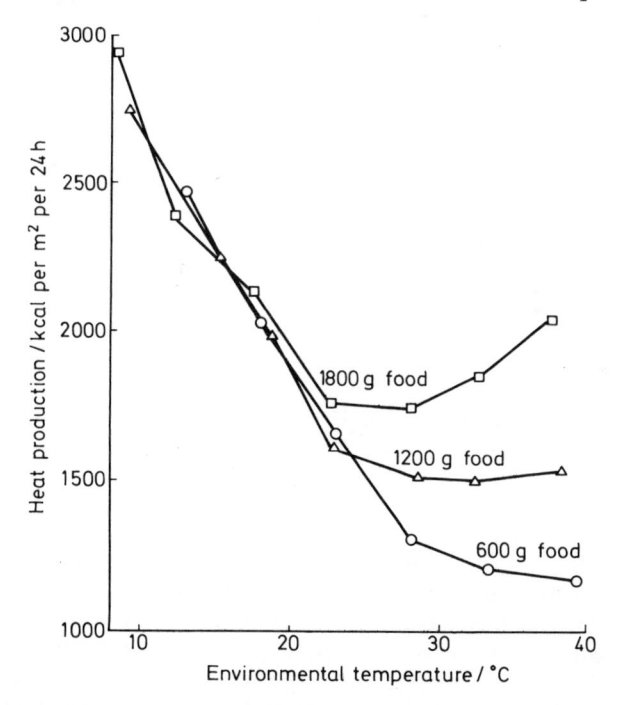

Figure 1.1 Heat loss in closely clipped sheep in relation to environmental temperature and feeding level (Graham et al., *1959)*

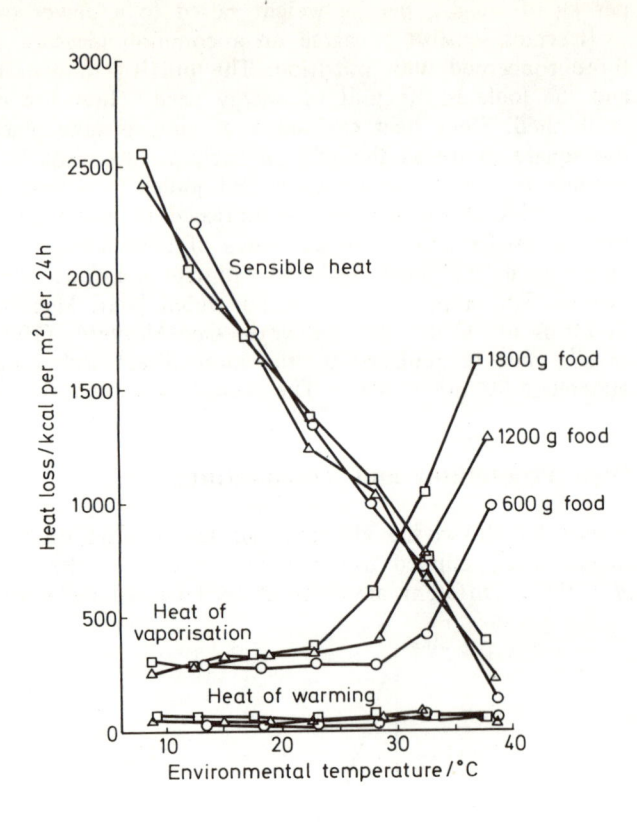

Figure 1.2 Partition of heat losses of the animal shown in Figure 1.1

but only above a temperature of about 25°C. Below this, heat production was independent of the food supply and entirely determined by the environment. *Figure 1.2* shows that this effect was due entirely to the invariant nature of the rate of heat loss by sensible means, that is by radiation, convection and conduction, which increased by a constant amount with each fall in environmental temperature by 1°. In addition, the heat lost by vaporising moisture from the skin and respiratory passages was invariant at low temperatures. This same general relationship is seen in all species. *Figure 1.3*, for example, summarizes work with human infants.

The lower critical temperature of an animal is defined as that temperature below which heat production increases in response to a fall in environmental temperature, and can be expressed as:

$$T_C = T_R - \frac{1}{c}(H-E)$$

critical temperature = rectal temperature − reciprocal of the negative slope of the relationship between sensible heat loss and temperature × {Heat production − Minimal heat loss by evaporation}

$\frac{1}{c}$ is in effect 'total insulation', and is the resistance to heat flow placed between the deep body and the air. At temperatures below the critical

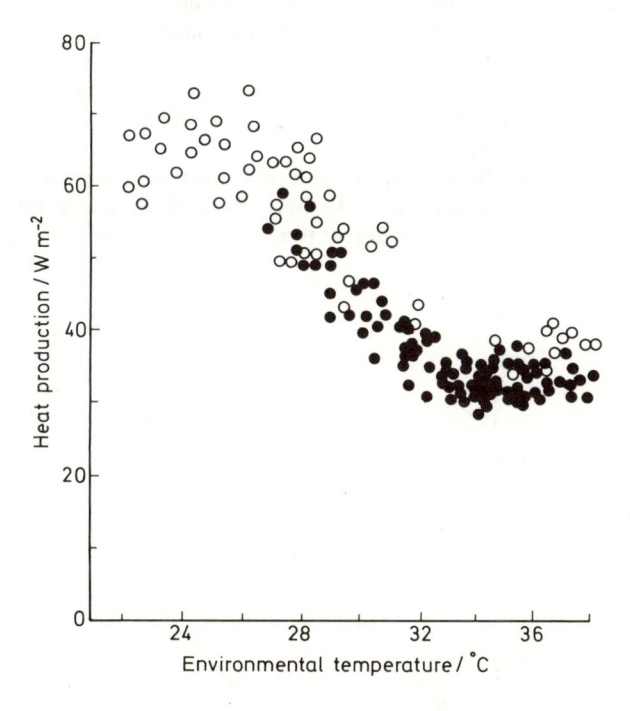

Figure 1.3 Environmental temperature and heat production in human infants

temperature, heat production increases by a constant amount 'c' for every °C fall in temperature, i.e. the reciprocal of the insulation is the heat loss per degree. Equally, it is a measure of the amount of metabolisable energy which should be given to an animal as food for every degree that its environment is lower than the lower critical temperature to maintain productivity at the level which appertains at warmer temperatures.

6 *Environmental factors – influence on nutrition of farm livestock*

The relationship between temperature and food can be seen in more practical terms in an experiment with pigs by Fuller and Boyne (1971) as shown in *Figure 1.4*. To maintain the same rate of gain of 700 g per day at 5°. as at 23° pigs had to be given more food; a fall of 1°C being equivalent to an increase of 2.3 g feed per kg $W^{0.73}$ (W = weight). For a 50 kg pig and a house temperature of 13° rather than 23° this is equivalent to 400 g feed per day. Verstegen *et al.* (1973) and Close, Mount and Start (1971) found in their experiments values of 250–300 g for a 10°C fall. Verstegen and van der Hel (1974) found 300 g per °C per day in somewhat smaller pigs. None of these are small amounts of feed, but exact determination depends on the precise nature of the total insulation of the animal and its critical temperature. To estimate these one must delve deeper into the physics of heat loss.

Figure 1.5 shows the relationships between an animal producing heat and its environment. Heat is produced centrally and is convected to the surface of the body in blood. The animal can control this flow by opening or closing capillary networks in the skin. At the skin surface some heat is used to vaporise moisture and the rest is transferred through the coat surface where it is transferred to the environment by radiation and by convection.

We can distinguish two terms, ideally three terms, making up the total insulation or resistance to heat flow from the animal: the tissue insulation of the animal, and the external insulation which includes that of the coat

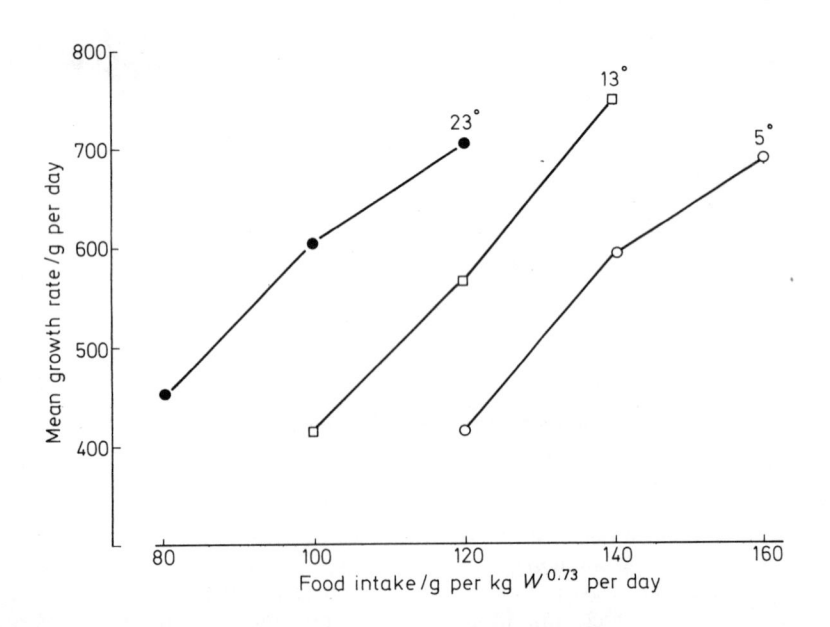

Figure 1.4 Food intake and gain in pigs kept at different temperatures (Fuller and Boyne, 1971)

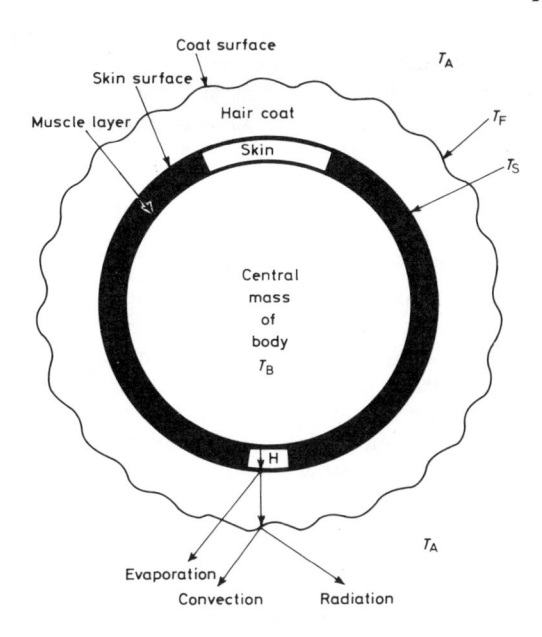

Figure 1.5 Diagrammatic representation of heat flow in an animal

and that of the boundary layer of the coat surface, or in a hairless animal, of the skin surface. These insulations are defined in the following way:

Tissue insulation, $I_T = \dfrac{T_R - T_S}{H \text{ m}^{-2} \text{d}^{-1}}$

(temperature gradient rectal to skin divided by rate of heat loss per m² per day)

External insulation, $I_E = \dfrac{T_S - T_A}{(H-E) \text{ m}^{-2} \text{d}^{-1}}$

(temperature gradient skin to air divided by the rate of sensible heat loss per m² per day)

'Fleece insulation' and 'air insulation', I_F and I_A, together sum to give external insulation.

It should be noted that the external insulation is defined in relation to the temperature of the air. Air temperature is not the sole index of the 'coldness' of an environment. We have to consider air movement, precipitation which alters the coat insulation, and the short-wave and long-wave radiation environment. Clear skies at night cause an increase in sensible loss due to reduced incoming long-wave radiation; during sunshine there is a decrease in loss of heat as a result of increased incoming short-wave radiation. Obviously a precise analysis must involve details of the physics of each of these energy flows — and in cases mass flows — but in a conventional treatment in terms of an external insulation it has to be appreciated that insulation as defined above will vary with wind velocity, the radiation environment and the amount of rain or moisture impinging on and wetting the coat.

Tissue Insulation

In the individual animal the tissue insulation in cold is very constant and invariant with environmental conditions except that below about 5°C it falls by about 10—20% (Webster and Blaxter, 1966; Webster and Young, 1970). The reason for this fall is that in cold environments the temperatures of tissues without underlying muscle capable of shivering (ears, shanks, feet and the terminal part of the tail) fall to temperatures about 2—4° above that of the air. To prevent freezing of these tissues at sub-zero temperatures a reflex vasodilation takes place, to allow warm blood to enter. Characteristically the temperature of such tissues exhibits a sawtooth sequence in time, and obviously the insulation is reduced when this occurs. Young animals have lower tissue insulations than older ones (Gonzalez-Jimenez and Blaxter, 1962; Holmes, 1970) and mature values appear to be reached in cattle in about a month. Webster, Chlumecky and Young (1970), however, in Canada have suggested that tissue insulation increases with weight to give values about twice those noted in Scotland. These authors recognised that this could well have arisen as a result of habituation to continued cold, and for UK conditions the lower values are preferred. In addition, increased body fatness, particularly in young pigs (Ingram, 1964) and in sows (Holmes and McLean, 1974) increases tissue insulation probably by a maximum of 40%. The values in different species are summarised in *Table 1.1*.

Table 1.1 Tissue insulations (°C m^2 d MJ^{-1}) in different species and classes of animal, measured below the critical temperature but above that which induces vasodilation to prevent tissue damage

Species and class	Tissue insulation	Reference
Newborn calf	0.65	Gonzalez-Jimenez and Blaxter, 1962
Month-old calf	1.62	Gonzalez-Jimenez and Blaxter, 1962
Adult steer	1.59 ± 0.30	Blaxter and Wainman, 1964
Adult steer	1.22 and 1.45	Blaxter and Wainman, 1961
Adult steer (500 kg)	3.47	Webster, Chlumecky and Young, 1970
Newborn pig	0.24	Mount, 1959
Adult pig	1.67	Irving, Peyton and Monson, 1956
Adult sheep	1.32	Joyce and Blaxter, 1964
Adult sheep	1.35	Webster and Blaxter, 1966

External insulation

If the external insulation of an animal is measured in relatively still air with the radiant environmental temperature close to ambient temperature and regressed on coat depth measured with a probe, the slope is a measure of the insulation of the coat per unit depth and the intercept term is a measure of the convective and radiative insulation of the interface at the coat surface. *Table 1.2* summarises the values obtained for

Table 1.2 External insulation ($°C\ m^2\ d\ MJ^{-1}$) of species and classes of animal in conditions of minimal air movement and when radiant temperature is close to air temperature in relation to coat depth F (cm). The intercept approximates 'air insulation' and the slope the reciprocal of the thermal conductivity of the coat.

Species and class	$I_E = a + bF$	Reference
Young calf – coat erect	$I_E = 1.43 + 0.79F$	Gonzalez-Jimenez and Blaxter, 1962
Adult steer	$I_E = 1.64 + 0.88F$	Blaxter and Wainman, 1964
Adult cattle	$I_E = 2.33 + 0.57F$	Webster, Chlumecky and Young, 1970
Mature sheep	$I_E = 1.35 + 0.141F$	Joyce, Blaxter and Park, 1966
Young pig	$I_E = 1.36 + 0.035F$	Mount, 1964
Man (naked)	$I_E = 1.39$	Bazett, 1949

'air insulation' and 'fleece insulation' estimated in this way. It is remarkable how little interspecies variation is present in the intercept term, the high value obtained by Webster, Chlumecky and Young (1970) being anomalous. Equally, the slopes show what might be expected. A fleece on the sheep is a better insulator per cm than a cattle coat, and the sparse hair of a pig is very inferior. If one compares estimates of the insulation of 20 mm coats in cattle, the value obtained by Blaxter and Wainman (1964) is 3.40 and by Webster, Chlumecky and Young (1970) 3.47, while in calves the value is 3.01, suggesting that in the working range these regressions do not necessarily give grossly discordant results.

Wind destroys the insulation of coats and augments the surface heat exchange. Methods have been devised by Joyce, Blaxter and Park (1966) and by Webster, Chlumecky and Young (1970) to predict these, the former one generally and the latter specifically. Values for sheep are summarised in *Table 1.3* and for cattle in *Table 1.4*. The latter equation predicts external insulation of cattle in wind to be considerably greater than that of sheep. A comparison of direct determinations of heat loss in wind made in New Zealand (Holmes and McLean, 1974) and in Scotland (Blaxter and Wainman, 1964) (*Figure 1.6*), with the prediction equation of Joyce, Blaxter and Park (1966) suggests that for temperate conditions this is probably more suitable.

Rain affects external insulation but the effects have not been quantified within the scheme. From Alexander's work on wet lambs, not only is insulation reduced, but additional heat is required to dry out the coat.

Table 1.3 External insulation ($°C\ m^2\ d\ MJ^{-1}$) of sheep in relation to wind velocity and fleece depth*

Wind velocity (miles per hour)	External Insulation					
	10mm**	20mm**	30mm**	40mm**	50mm**	60mm**
0.4	2.49	3.58	4.60	5.58	6.51	7.39
1.0	2.20	3.23	4.20	5.12	5.99	6.83
2.0	1.95	2.90	3.81	4.66	5.48	6.25
4.0	1.66	2.51	3.31	4.08	4.80	5.49
8.0	1.34	2.04	2.70	3.32	3.91	4.48
16.0	0.97	1.44	1.89	2.32	2.72	3.11

*Estimated from the equation of Joyce, Blaxter and Park (1966), viz.,

$$I_E\ (°C\ m^2\ d\ Mcal^{-1}) = \frac{r}{r+F} \cdot \frac{1}{0.115 + 0.099\sqrt{v}} - r\ \log_e \frac{r+F}{r}\ (z - 0.09\sqrt{v})$$

where r is the radius of a sheep (150 mm)
F is the depth of fleece in mm
v is air velocity in miles per hour
z is the thermal insulation of 1 mm fleece (0.59 $°C\ m^2\ d\ Mcal^{-1}$)

The values in the table have been converted to MJ by dividing by 4.184
**Mean depth of fleece measured with a probe

Table 1.4 Alternative estimates of the external insulation ($°C\ m^2\ d\ MJ^{-1}$) of cattle in relation to wind velocity and coat depth

Wind velocity (miles per h)	Results from general equation[1]			Results from regression analysis[2]		
	10mm*	20mm*	30mm*	10mm*	20mm*	30mm*
0.4	2.26	3.16	4.04	3.01	3.58	4.15
1.0	1.96	2.79	3.60	2.79	3.36	3.93
2.0	1.70	2.44	3.17	2.53	3.31	3.68
4.0	1.41	2.03	2.65	2.18	2.75	3.32
8.0	1.08	1.53	1.98	1.68	2.25	2.82
16.0	0.69	0.91	1.13	0.96	1.53	2.10

[1] From the general equation inserting 0.46 $°C\ m^2\ Mcal^{-1}$ for the insulation of cattle coats and a value of 400mm for r (see Table 1.3).
[2] From regression analysis of observations on Canadian cattle (Webster, Chlumecky and Young, 1970)
*Coat depth

Table 1.5 gives the results of experiments with cattle and sheep which show that rain and wetting of the coat increase heat loss and the effects of wind and rain are additive in increasing this loss. The results of Webster and Park (1967) indicate that in fleeced sheep external insulation is reduced by about 2 $°C\ m^2\ d\ MJ^{-1}$ by continuous rain. Similar considerations apply to changes in the long-wave radiation environment. The

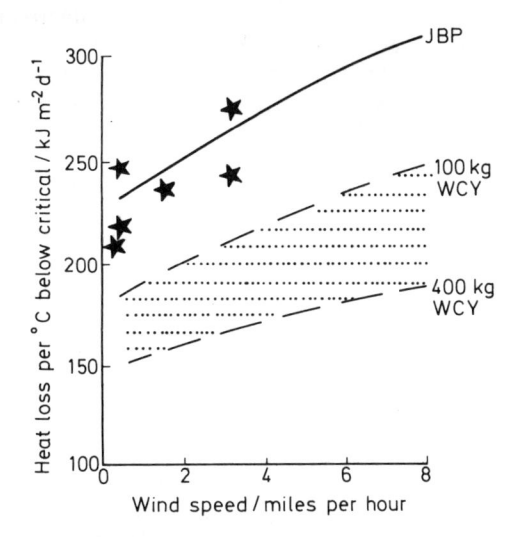

Figure 1.6 Heat losses below the critical temperature of cattle as predicted by two methods and (★) observed in 24 hour calorimetric experiments. (JBP predicted heat losses according to Joyce, Blaxter and Park, 1966; WCY predicted heat losses according to Webster et al., 1970.)

Table 1.5 The effect of wind and rain separately and together on the daily heat production of calves, adult cattle and sheep. The animals were at environmental temperatures below their critical ones

Animal and class	Conditions Wind speed (miles/h)	Duration of rain in 24h	Metabolism in the cold (MJ d^{-1})	Increase MJ d^{-1} due to Wind	Rain	Both together	Reference
Calves	3.5	4h	9.9	1.0	0.8	2.3	1.
Cattle – close clipped	1.6	7h	54.5	4.6	4.0	7.5	2.
Cattle – thick winter coat	1.6	7h	37.0	1.2	1.2	1.8	2.
Blackfaced sheep	10	continuous	4.2	1.4	3.5	5.1	3.
Suffolk sheep	10	continuous	5.3	0.9	3.3	5.7	3.
Newborn lamb (2 kg)	12.5	wetted at birth	0.86	0.86	0.86	1.72	4.

1. Holmes and McLean (1975)
2. Blaxter and Wainman (1965)
3. Blaxter, Clapperton and Wainman (1966)
4. Computed from Alexander's (1974) summary of his extensive and critical work, that thermoneutral metabolism is 70 W, and wind at 5.5 m s^{-1} and a wet coat increase metabolism by 70 W each and the effects are additive. *See also* Alexander (1962)

12 *Environmental factors – influence on nutrition of farm livestock*

effects have been measured both by calorimetry (Mount, 1964; Joyce, Blaxter and Park, 1966) and by the device of using an artificial animal to integrate the effect (Webster, 1971). The effect is to reduce the magnitude of the first term in the regressions in *Table 1.2*. It thus must be kept in mind that to use expressions for predicting external insulation that take into account only coat depth and air velocity will provide values which will have to be revised *downwards* if it is raining or the animals are wet or if at night the skies are clear. The effects of solar radiation during the day have been described and quantified (Clapperton, Joyce and Blaxter, 1965). They obviously effectively increase insulation measured in this way.

Estimation of Critical Temperature and Food Need

From the information given on insulations in *Tables 1.1, 1.2, 1.3* and *1.4*, together with the additional information that the minimal loss of heat by vaporising moisture is 1.5 MJ m^{-2} d^{-1} in cattle (Blaxter and Wainman, 1961), 1.3 MJ m^{-2} d^{-1} in sheep (Joyce and Blaxter, 1964) an 0.7 MJ m^{-2} d^{-1} in pigs (Ingram, 1964), critical temperature, T_C, can be estimated from the equation given below:

$$T_C = T_R + EI_E - H(I_T + I_E)$$

where T_R is rectal temperature (39°), E the minimal loss of heat by vaporising moisture, H the heat arising from metabolism, I_T tissue insulation and I_E external insulation.

An example of the method used and the integral part that tables of animal requirements play in manipulating this equation is given below.

Consider a heifer with a 1 cm coat which it might have in early autumn weighing 200 kg and fed to gain 0.5 kg d^{-1}. From either MAFF (1975) or ARC (1965) tables of metabolisable energy (ME) requirement, heat production can be estimated. This heat consists of the maintenance ME requirement plus (1–k_f) times the production ME requirement.

From ARC requirements (*Table 6.27*) maintenance is 33.4 MJ d^{-1} and total requirement 44.4 MJ d^{-1}. Total heat is therefore 33.4 + [(44.4 – 33.4) × 0.49]= 38.8 MJ d^{-1}, since for the diet concerned the efficiency of utilisation of metabolisable energy is 51%. A beast of 200 kg has a surface area of 3.0 m^2 (*see* Appendix). Heat production on a surface area basis is thus 38.8/3 = 12.9 MJ m^{-2} d^{-1}.

The tissue insulation of this animal is 1.59 units (*Table 1.1*) and if the environment is mildly windy – say 4 miles per h – its external insulation will be 1.41 units (*Table 1.4*). Then, by substitution of these values in the equation $T_C = T_R + EI_E - H(I_T + I_E)$:

$$T_C = 39.0 + (1.5 \times 1.41) - 12.9 (1.59 + 1.41)$$

$$T_C = +2.4°C.$$

If the hair coat increases during the winter from 1 cm to a good thick 3 cm, then external insulation would be 2.65 and T_C not +2.4°C but −11.7°C. If food intake was increased to permit a gain of 0.75 kg, this would result in a heat production of 13.9 MJ m^{-2} d^{-1} and the critical temperature would fall still further to −16.0°C.

The additional food required to maintain gain at a prescribed rate in an environment colder than the critical temperature is given by the relationship:

$$\begin{array}{l}\text{additional food energy} \\ \text{required (MJ m}^{-2}\text{ d}^{-1}) \\ \text{to prevent reduction} \\ \text{of gain when} \\ \text{temperature falls to} \\ T_X \text{ (below } T_C)\end{array} = (T_C - T_X) \bigg/ \left(\frac{HI_T}{H-E} + I_E \right)$$

The denominator in this expression is the food per m^2 d per degree centigrade that the environment is below the critical temperature.

For the animal fed to gain 0.5 kg with a 1 cm coat this would be:

$$\text{MJ m}^{-2}\text{ d}^{-1}\text{ °C}^{-1} = \left(\frac{12.9}{12.9 - 1.5} \times 1.59 + 1.41 \right)^{-1}$$

$$= \frac{1}{1.13 \times 1.59 + 1.41}$$

$$= 0.321 \text{ MJ m}^{-2}\text{ d}^{-1}\text{ °C}^{-1}$$

Since the animal has a surface area of 3 m^2 this amounts to 0.96 MJ d^{-1} °C^{-1}.

Thus to combat a temperature 10°C below the critical, 9.6 MJ feed have to be provided each day, equivalent to 0.8 kg barley.

If the animal has a thicker coat and a lower critical temperature as in the second illustration, then the additional feed energy required for every degree that temperature falls below the critical is not 0.32 MJ m^{-2} d^{-1} °C^{-1} but much less, namely 0.22 MJ m^{-2} d^{-1} °C^{-1}. Generally, the lower the critical temperature of an animal the smaller is the amount of feed needed to keep the animal warm without a diminution of its productivity when temperature falls below the critical temperature.

Discussion

From the data presented the critical temperatures of stock can be predicted for various conditions involving wind velocity. More sophisticated approaches allow estimation of the effects of long- and short-wave radiation and largely empirical approaches allow the effects of rain to be approximated. It has to be kept in mind that the calculations given above apply to dry conditions in which the sky is overcast so that the radiation temperature of the sky is close to that of the air. The calculations certainly apply to conditions within buildings or simple shelters.

There is evidence (Slee, 1971; Webster, 1974) that farm animals adapt to cold. One adaptation is that they eat more and increase their heat production; it might seem that they appreciate the value of food in combating cold. More important, they increase their insulation by developing very thick winter coats. This is seen not only in cattle but also in pigs kept continuously in the cold. In addition, in long-continued severe cold there is evidence of an habituation to cold which in effect appears to increase total insulation in other ways. Certainly the storm of last April probably had such severe effects because it followed an open mild winter. Indeed, a farmer's remark about his losses was that the sheep were 'green'. The equations should allow prediction of food needs and critical temperatures in such circumstances; coat depth can be measured while feed intake can also be determined and the heat produced by the animal calculated.

These considerations of the direct effects of environment on the animal and their interaction with the feed supply are of value in the design of buildings and shelters, the estimation of the likely return from investment in insulation, the design of supplementary heating plants as well as in the forward planning of feed supplies. They allow the environment to be assessed in terms of its effect on the feed required to maintain production. Much more work is required to make the approach exact, to enable it to integrate the hour-to-hour variability in the outdoor environment in our country where every passing cloud modifies the environmental demand for heat and to resolve and refine the measurements. Even so, research has already provided sufficient information to enable many practical questions about the effects of the climatic environment to be answered.

References

Alexander, G. (1962). *Aust. J. agric. Res.*, **13**, 82
Alexander, G. (1974). In *Heat Loss from Animals and Man*, p.205. Ed. by J.L. Monteith and L.E. Mount. London; Butterworths
ARC (1965). *The Nutrient Requirements of Farm Livestock. No. 2. Ruminants.* London
Bazett, H.C. (1949). In *Physiology of Heat Regulation*, p. 109. Ed. by L.H. Newburgh. Philadelphia; Saunders
Blaxter, K.L. and Wainman, F.W. (1961). *J. agric. Sci., Camb.*, **56**, 81
Blaxter, K.L. and Wainman, F.W. (1964). *J. agric. Sci., Camb.*, **62**, 207
Blaxter, K.L. and Wainman, F.W. (1965). Unpublished work carried out at the Hannah Institute
Blaxter, K.L., Clapperton, J.L. and Wainman, F.W. (1966). *Br. J. Nutr.*, **20**, 283
Clapperton, J.L., Joyce, J.P. and Blaxter, K.L. (1965). *J. agric. Sci., Camb.*, **64**, 37

Fuller, M.F. and Boyne, A.W. (1971). *Br. J. Nutr.*, **25**, 259
Gonzalez-Jimenez, E. and Blaxter, K.L. (1962). *Br. J. Nutr.*, **16**, 199
Graham, N. McC., Blaxter, K.L., Wainman, F.W. and Armstrong, D.G. (1959). *J. agric. Sci., Camb.*, **52**, 13
Holmes, C.W. (1970). *Anim. Prod.*, **12**, 493
Holmes, C.W. and McLean, N.R. (1974). *Anim. Prod.*, **19**, 1
Holmes, C.W. and McLean, N.R. (1975). *N.Z. Jl agric. Res.*, **18**, 277
Ingram, D.L. (1964). *Res. vet. Sci.*, **5**, 357
Irving, L., Peyton, L.J. and Monson, M. (1956). *J. appl. Physiol.*, **9**, 421
Joyce, J.P. and Blaxter, K.L. (1964). *Br. J. Nutr.*, **18**, 5
Joyce, J.P., Blaxter, K.L. and Park, C. (1966). *Res. vet. Sci.*, **7**, 342
MAFF (1975). *Technical Bulletin No. 33.* London; HMSO
Monteith, J.L. (1973). *Principles of Environmental Physics.* London; Edward Arnold
Monteith, J.L. and Mount, L.E. (1974). *Heat Loss from Animals and Man.* London; Butterworths
Mount, L.E. (1959). *J. Physiol., Lond.*, **147**, 333
Mount, L.E. (1964). *J. Physiol., Lond.*, **173**, 96
Slee, J. (1971). *Proc. Nutr. Soc.*, **30**, 215
Verstegen, M.W.A., Close, W.H., Start, I.B. and Mount, L.E. (1973). *Br. J. Nutr.*, **30**, 21
Verstegen, M.W.A. and van der Hel, W. (1974). *Anim. Prod.*, **18**, 1
Webster, A.J.F. (1971). *J. appl. Physiol.*, **30**, 684
Webster, A.J.F. (1974). In *Heat Loss from Animals and Man*, p.205. Ed. by J.L. Monteith and L.E. Mount. London; Butterworths
Webster, A.J.F. and Blaxter, K.L. (1966). *Res. vet. Sci.*, **7**, 466
Webster, A.J.F. and Park, C. (1967). *Anim. Prod.*, **9**, 483
Webster, A.J.F. and Young, B.A. (1970). See Monteith and Mount, 1974
Webster, A.J.F., Chlumecky, J. and Young, B.A. (1970). *Can. J. Anim. Sci.*, **50**, 89

Appendix

HEAT PRODUCTION PER UNIT TIME

From	To	Factor
Mcal d^{-1}	MJ d^{-1}	Multiply by 4.184
Mcal d^{-1}	Watts	Multiply by 48.4

INSULATION TERMS (temperature gradient divided by heat flow per unit area)

°C m^2 d Mcal^{-1}	°C m^2 d MJ^{-1}	divide by 4.184
°C m^2 d Mcal^{-1}	°C m^2 W^{-1}	divide by 48.4

16 *Environmental factors – influence on nutrition of farm livestock*

BODYWEIGHT, SURFACE AREA AND METABOLIC BODY SIZE*

Bodyweight (kg)	Surface area (m^2)	Metabolic size ($W^{0.75}$)
1	0.09	1.0
2	0.14	1.7
15	0.26	3.3
25	0.76	11
50	1.2	19
100	1.9	32
200	3.0	53
400	4.9	89
600	6.4	121

*These are calculated from a general formula A = $0.09W^{2/3}$, the Meeh formula. It is not absolutely precise for a particular species, indeed within species the power of weight with which area varies is usually in the range 0.5–0.6 with the coefficient correspondingly higher (0.12–0.15). For approximate calculations the general formula is sufficiently accurate.

2

NUTRITION–ENVIRONMENT INTERACTIONS IN POULTRY

A.H. SYKES
Wye College, University of London

The scope of this review has been restricted partly by selection of only certain aspects of the subject but mainly by limitations of the available data.

The term environment will be restricted to ambient temperature, thus omitting daylength, stocking density and other aspects of the environment. Moreover, climates will be dealt with in which the only variable is taken to be the dry bulb thermometer reading. Variations in other climatic factors such as humidity and wind speed have been ignored since until there is an effective temperature scale for the fowl there is no means of evaluating the importance of these factors; they are seldom recorded in the experiments to be discussed. A further limitation is that only constant, as distinct from fluctuating, temperatures are considered. No two authors have used the same combination of fluctuating temperatures and time periods and therefore comparisons between experiments becomes very difficult and mean temperature alone does not indicate the severity of a climate.

Defining climate solely in terms of constant ambient temperature does not make it easier to apply the results to natural climates, which are very complex situations, but makes it possible to determine whether, in simplified and reproducible experimental conditions, there are any changes in nutrient requirements.

This leads to the question, 'requirements for what?' to which the answer must be, 'for maximum production of eggs, meat or chicks hatched as limited only by non-nutritional factors such as genetics, disease or stress'. This may appear obvious but it is not always appreciated that requirements are not absolute but are related to level of production, a point well illustrated by Fisher (1976) in relation to vitamin A requirements. Thus at any given ambient temperature (T_A) the percentage egg production or rate of growth should be specified.

Requirements should be stated in terms of intake per unit time; for most practical purposes intake per day is suitable since the day is a natural unit within which physiological cycles of feeding and other metabolic activities occur but shorter or longer periods could be used especially if there are body reserves of nutrients which require replenishing more or less frequently. The dietary inclusion rate of a nutrient is, of course, less satisfactory as a statement of requirements and is positively misleading when food intake falls below the expected norm, as illustrated by Bray and Gesell (1961) in relation to protein requirements. The converse, i.e. a dietary excess when food intake is above normal, presumably occurs but has not been described since the consequences are unlikely to be detrimental to production. A somewhat similar situation arises when a dietary deficiency, e.g. of lysine, at one T_A is ameliorated, as judged by increased growth rate, when the T_A is reduced and food intake increases (March and Biely, 1972).

Much as one would like to see tables of recommended daily nutrient intake for a wide range of T_A our knowledge at present is too limited to do more than make rough approximations in respect of one or two nutrients. The situation is still very much as described by Emmans (1974) in his discussion of dietary energy. Many publications cannot be considered because they cover only a very narrow temperature range, do not provide details of diets used or have other inadequacies. Ideally a range of daily nutrient intakes should be provided, thus allowing response curves to be constructed at each of several T_A within the range normally experienced, say $-5°$ to $+35°C$. There is an obvious tendency to study only T_A likely to be found in the UK or temperate zones generally but industrialised poultry production is extending its climatic range including, for example, such extreme climates as Saudi Arabia and the Sudan and it is valuable to know the nutritional responses to climate in order to determine what other factors impose limits on production. There is surprisingly little information on the nutritional effects of extreme cold although there must be some experience of this from parts of Canada or Northern Europe.

Finally, because of limitations of the data it has been necessary to exclude from consideration breeding stock and broilers and the discussion is concerned solely with the effects of constant ambient temperatures on the nutrient requirements of laying poultry.

The nutrients to be considered are energy, protein, water and two vitamins. Throughout, values are quoted per bird or on a calculated bodyweight of 1.5 kg; the use of unit weight or powers reduces variation but makes the data less comprehensible in practical terms.

Energy Requirements

Unlike most other nutrients, except water, energy is usually supplied *ad libitum* and it is assumed that requirements are satisfied by voluntary intake. There do not appear to be any trials at different T_A giving response curves to a range of daily intakes although some variation of intake is sometimes achieved by the use of different energy concentration. The requirement has therefore to be estimated from the graph relating metabolisable energy (ME) intake to T_A. For this purpose data from 9 trials have been used. Plotting some of these trials separately (*Figure 2.1*) showed a considerable range of intakes at any given T_A, as might be expected from the varying conditions, strains and levels of production.

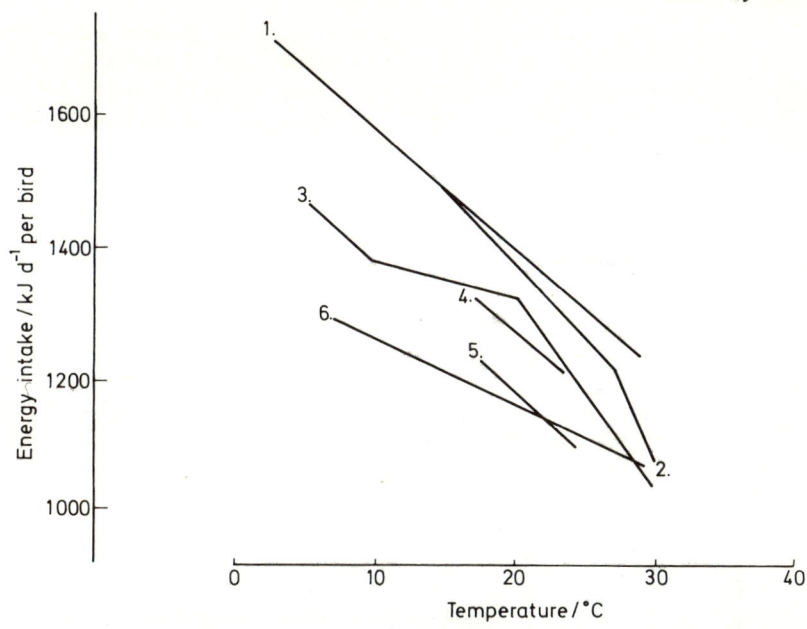

Figure 2.1 The intake of metabolisable energy in relation to ambient temperature. Examples of different trials. 1. Ota (1960); 2. Morris (1975); 3. van Es et al. (1973); 4. Emmans, Charles and Dun (1975) medium strain; 5. Emmans, Charles and Dun (1975) light strain; 6. Davis, Hassan and Sykes (1973). Values expressed in kJ per bird per day.

The main trend, a lower ME requirement with increasing T_A, is seen better when mean intakes were plotted at 5°C intervals (*Figure 2.2*); alternatively all the data were used to construct a regression line (*Figure 2.2*) which is close to the mean but does not show the greater reduction of intake which takes place at 30–35°C. At 20°C, the middle of the range, the ME intake is 1297 kJ per d with a change of 21 kJ per d per °C. This mean value derives from small-scale energy balance trials on carefully selected birds (Davis, Hassan and Sykes, 1973), calorimetric estimations over 4 day periods (van Es et al., 1973) and longer term laying trials under more practical conditions (Emmans, Charles and Dun, 1975; Smith, 1973; Ota, 1960; Ito et al., 1970; Payne, 1967; Morris, 1975; Nasir, 1975). It is the most precise measurement available of a change in nutrient requirement brought about by ambient temperature.

A number of authors have constructed formulae which predict the energy intake required for a given bodyweight and level of production. That of Byerly as modified by Combs (1962) gives a correction factor for season; the latest (ARC, 1975), refers only to average UK conditions; that of Emmans (1974) specifically refers to T_A. These three equations have been used on data from birds kept at 15°C and 30°C in three recent trials with the results shown in *Table 2.1*.

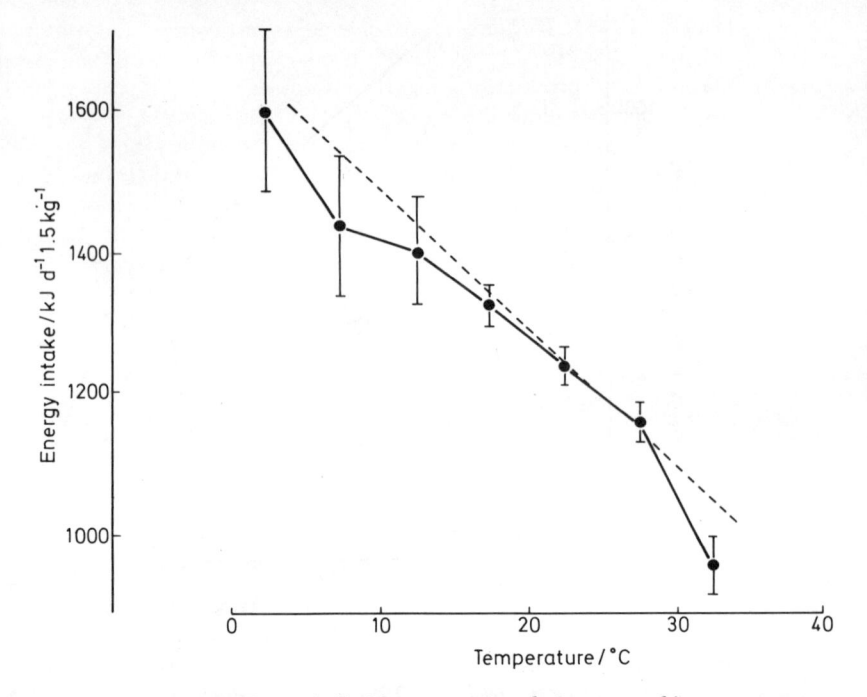

20 *Nutrition–environment interactions in poultry*

Figure 2.2 The intake of metabolisable energy in relation to ambient temperature; mean results from 9 trials at 5°C intervals expressed as kJ d^{-1} for a bird of 1.5 kg bodyweight. Dotted line: calculated regression. ME = 1690 − 20.1°C (r = 0.81)

Table 2.1 Actual and estimated energy intake at two temperatures (kJ d^{-1} per bird)

Reference T_A	Morris (1975) 15°C	30°C	Davis et al. (1973) 15°C	30°C	van Es et al. (1973) 15°C	30°C
Actual	1402	1088	1197	983	1368	1000
ARC	1464	1431	1536	1443	1481	1272
Byerly	1460	1418	1548	1410	1490	1146
Emmans	1385	1117	1452	1084	1494	962

The ARC and Byerly agree closely with each other at both temperatures but they overestimate particularly at 30°C whereas the Emmans equation gives good agreement especially at 30°C and it would be interesting to see how accurate it is in practical situations.

It is interesting to consider the relation between *ad lib.* ME intake, heat production and T_A. The concepts of a thermoneutral zone and a critical temperature for heat production are based upon the work of Rubner on the dog and have been applied to the fowl (Barott and

Pringle, 1946). If there is no change in efficiency and production remains constant then the curve of ME intake and T_A should be parallel to that of heat production, i.e. it should also exhibit a thermoneutral zone. But it appears that the curve of ME intake is straight over most of its temperature range with a tendency towards a parabola at the warmer end. However, more recent measurements of heat production on active, normally fed birds have also shown a linear response with T_A reaching a minimum at a new, and higher critical temperature above which the bird is in the hyperthermal zone and metabolism rises (*Figure 2.3*).

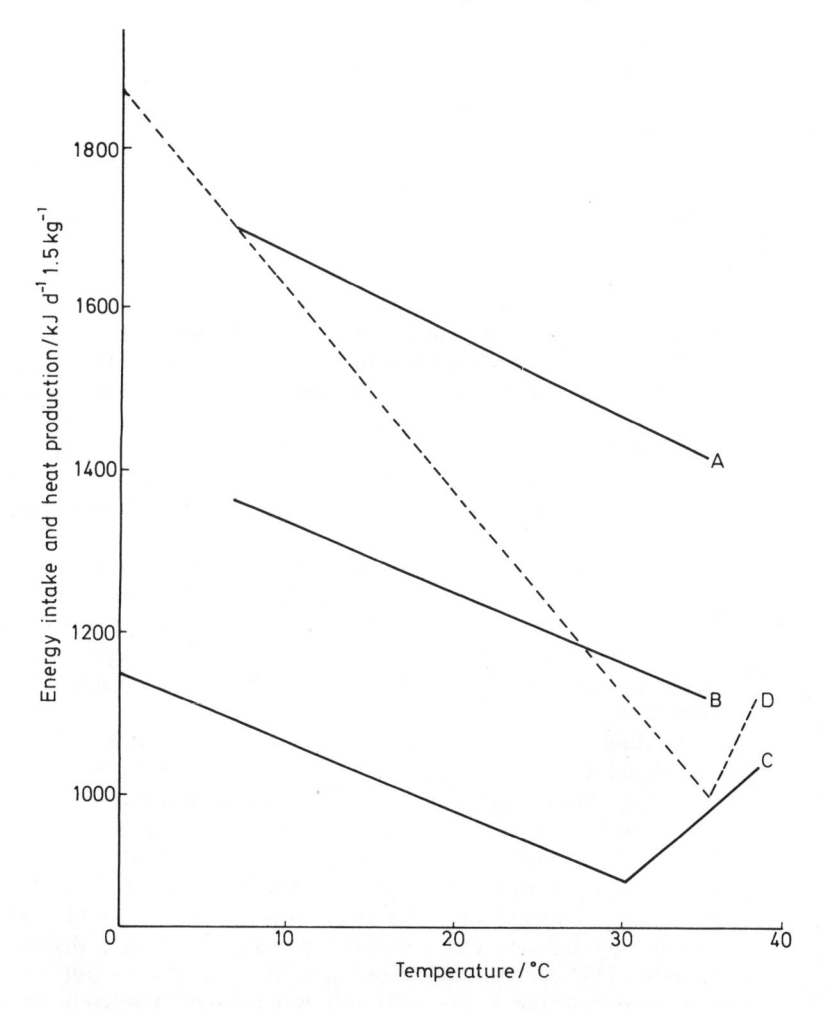

Figure 2.3 The intake of metabolisable energy and heat production in relation to ambient temperature. Line A, energy intake, and Line B, heat production, are from 3 week energy balances (Davis, Hassan and Sykes, 1973); Line C, heat production of well feathered hens, and Line D, heat production of poorly feathered hens, are from hourly rate of oxygen consumption (Richards, 1975). All values calculated to 1.5 kg bodyweight

This response has been shown by Romijn and Vreugdenhil (1969), Davis, Hassan and Sykes (1973) and van Es *et al.* (1973) and recently by Richards (1975) who measured oxygen uptake during 2 hours of exposure to a wide range of T_A. This curve is parallel with, but lower than, that of Davis, Hassan and Sykes (1973) but with a clear indication of a critical temperature around 32.5°C. He made similar measurements on very poorly feathered birds which showed much higher metabolic rates, a steeper slope and a somewhat higher critical temperature. The importance of insulation has always been recognised but there is little quantitative information available; it would be interesting to know the ME intake at different T_A of these poorly feathered birds. Results from cocks over a narrow range of T_A show that ME intake and heat production are about 50% higher after de-feathering (O'Neill, Balnave and Jackson, 1971). In the cold (5°C), heat production is more than doubled in poorly feathered birds (Romijn and Lukhorst, 1961). Clearly a complete description of factors affecting energy intake at a given T_A should include some measurement of insulation as well as bodyweight and rate of production.

The conventional explanation of the fall in ME intake with T_A is that the energy requirement for maintenance is reduced, since there is no reason to expect that the efficiency of egg formation changes with T_A. It follows that measurements of maintenance ME intake should be parallel to total ME intake but lower by the additional energy cost of egg production. The only attempt to measure maintenance energy at different T_A is that of Waring and Brown (1967) who found a value of 703 kJ per d per 1.5 kg at 22°C. The regression of −19.2 kJ per °C is similar to that found here for total ME intake which suggests that the above explanation is correct.

This partition of energy intake also assumes that the determination of ME remains unaffected by T_A and although one report (Smith, 1973) finds a lower ME at 38°C the majority find that ME remains constant or even increases slightly at warmer temperatures (van Es *et al.*, 1973; Davis, Hassan and Sykes, 1972; Olson, Sunde and Bird, 1972; Swain and Farrell, 1975).

Another aspect of environmental temperature which should be considered is the adaptation of ME intake to changes in T_A. The work of Davis, Hassan and Sykes (1972) showed that exposure to cold, 7°C from a temperature of 18°C, first resulted in a *reduction* in ME intake (*Figure 2.4*), possibly a stress response to the sudden change as observed previously by Campos, Wilcox and Shaffner (1962). This was followed by a sharp increase over the next week to bring the bird almost back into energy balance but without regaining the weight that had been lost initially. The energy intake per bird did not change but since bodyweight had been lost the intake per unit weight was effectively increased although only by about 5%, a reflection, perhaps, of the effective insulation of these birds. Exposure to a high temperature, 35°C, produced a similar sequence of events (*Figure 2.4*): an immediate reduction in ME intake followed by a gradual increase to a new energy balance but still at a greatly reduced level of both ME intake and bodyweight; the period of

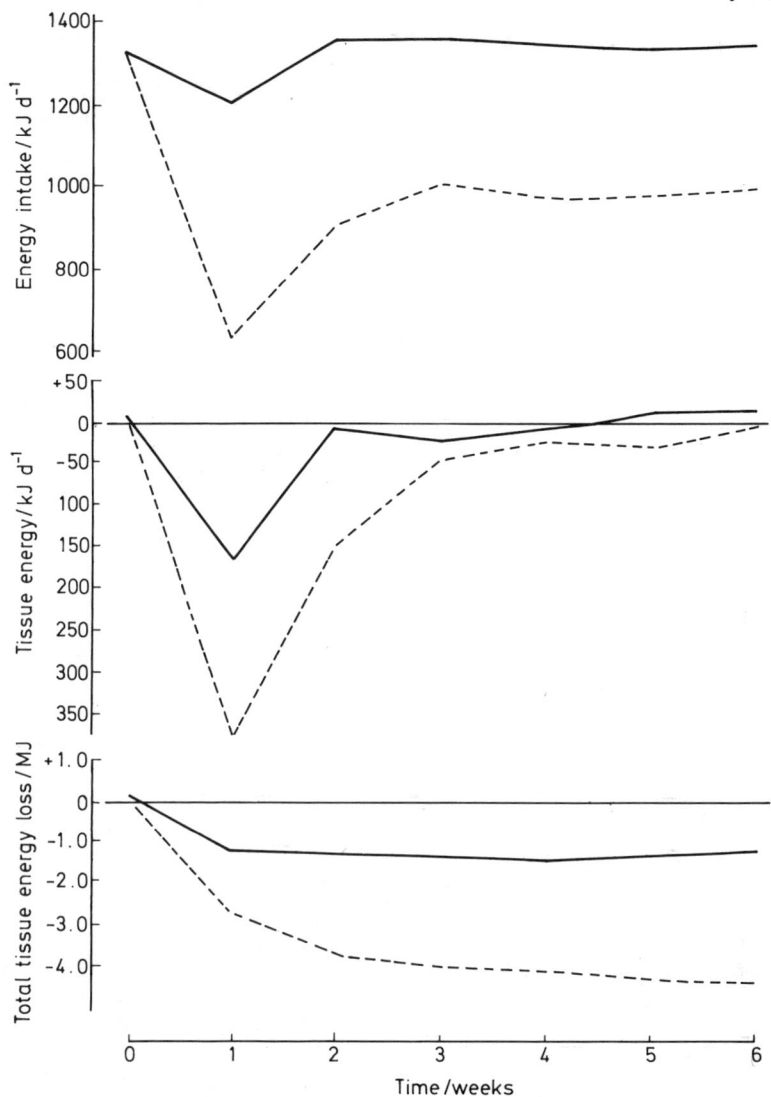

Figure 2.4 Adaptation of metabolisable energy intake and tissue energy to a change in ambient temperature. Solid lines = change from 18°C to 35°C; dotted lines = change from 18°C to 7°C. Upper figure = daily energy intake per bird; centre figure = daily loss of tissue energy from carcase analysis; lower figure = accumulative loss of tissue energy. Values expressed per bird. (from Davis, Hassan and Sykes, 1972)

adaptation was 3 to 4 weeks — appreciably longer than the time it takes for adaptation of fasting metabolic rate (Shannon and Brown, 1969) or rectal temperature (Hutchinson and Sykes, 1953). A feature of the response both to increase and decrease in T_A was the use of body reserves to help meet an immediate negative energy balance.

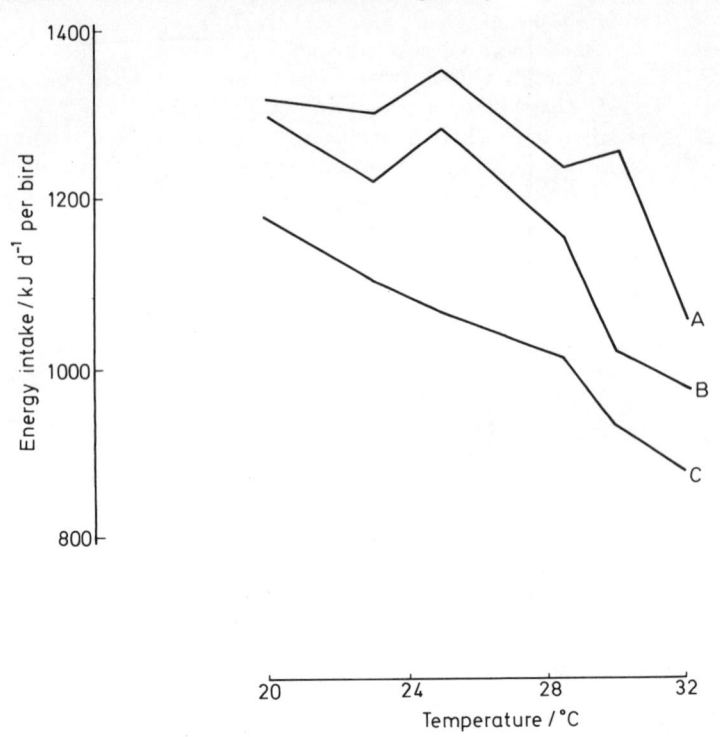

24 *Nutrition–environment interactions in poultry*

Figure 2.5 The intake of metabolisable energy in relation to ambient temperature as affected by dietary energy concentration. (A = 14.1 MJ per kg; B = 12.1 MJ per kg; C = 10.2 MJ per kg)

The slowness of the mechanism controlling food or energy intake to adapt can also be seen in several trials carried out at constant, warm temperatures (Mowbray and Sykes, 1971; Payne, 1967; Nasir, 1975) in which food intake did not change to compensate for alterations in dietary energy concentration. This is illustrated in *Figure 2.5* from Nasir (1975) and in *Table 2.2* from Payne (1967). In the latter, adaptation of

Table 2.2 Food and metabolisable energy intake in relation to dietary energy level and temperature (Payne, 1967)

Dietary energy (ME) (MJ kg^{-1})	Daily intake at 18°C		Daily intake at 30°C	
	Food (g)	ME (kJ)	Food (g)	ME (kJ)
11.97	127	1519	107	1280
12.80	118	1506	104	1330
13.60	112	1523	102	1389
14.43	106	1527	101	1452

food intake at 18°C was virtually complete whereas at 30°C food intake remained more or less constant and energy intake increased with dietary energy level. Under these circumstances what is the energy requirement, 1280 or 1452 kJ per d? And how far does voluntary intake reflect requirement? This observation has a bearing on the problems of tropical countries in which the poorer production, in egg number and egg size, is sometimes attributed to a low energy intake. In a sense this is true in that the same birds under temperate conditions would respond to an increase in the energy intake. But in the warm climate appetite is reduced, or the signals calling for an increased energy intake are over-ridden by other factors, possibly related to temperature regulation. A hen can tolerate 30°C without difficulty and has the ability to lose, by convection and evaporation, the additional heat load which results from ingesting, say, an additional 167 J per d. This was the difference between the highest and lowest intakes found by Payne at this temperature; an even greater range may be seen in the other references. Moreover, the lower the energy intake under warm conditions, the greater the loss of bodyweight as an energy source; why then does not the bird increase its intake to meet the deficit? The process of dietary adaptation has been shown to be slow, 3–4 weeks, but all the trials have been for periods of 12 or more weeks. Are there then other factors controlling food intake which become more important at higher temperatures? One possibility is the state of the water balance; it is well known that food consumption can be limited by water intake and it has also been shown that a water deficit leads to a reduction in metabolic rate (McFarlane, 1975). Eventually all nutritional deficiencies become energy deficiencies and to assert that tropical deterioration is an energy deficiency does nothing to identify the primary cause.

Protein Requirements

Bray and Gesell (1961) demonstrated decisively for the first time that egg production could be maintained at T_A 30°C provided a daily protein intake of about 15 g was ensured by appropriate dietary formulation (*Figure 2.6*). More recently Morris (1975) has shown that there is some response to increasing protein intakes up to 18 g d^{-1} but independently of T_A within the range of 15–27°C; at 30°C there was a lower egg output over the same range of intake but with no indication that dietary protein was limiting production. Although not conclusive, the results suggest that T_A does not increase or decrease the requirement for protein. Reid and Weber (1973) found that increasing the daily protein intake by stages from 12.7 to 20.5 g did not improve egg production at T_A 35°C although the level of production was low (44%) which prevents valid comparison with the controls (73%) at T_A 21°C.

With the maintenance metabolism falling as T_A rises it might be thought that protein requirements would also fall. There is clearly a need for further work over a wide range of temperature and protein or amino acid intake before this question can be answered.

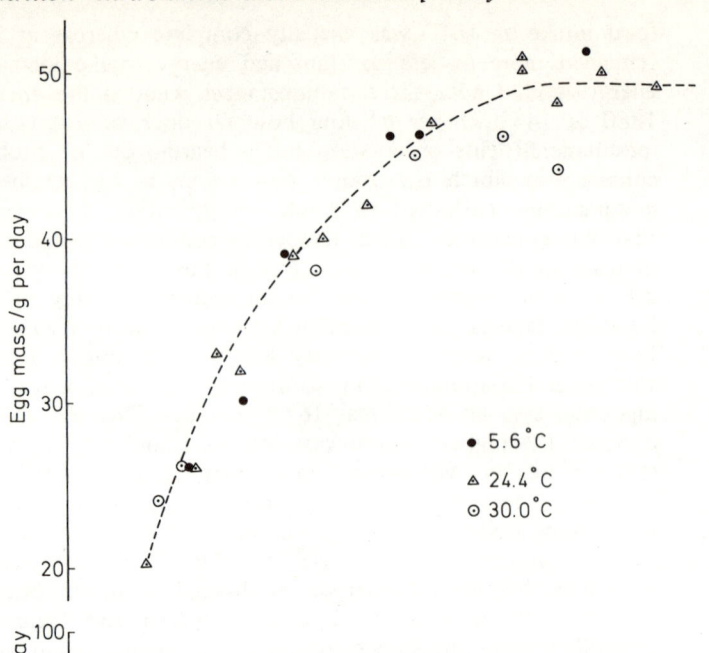

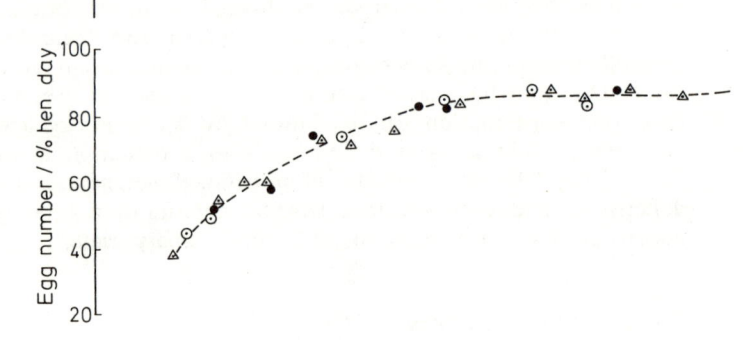

Figure 2.6 Egg production at different ambient temperatures as affected by protein intake. Upper = egg mass; lower = egg number. (Adapted from Bray and Gesell, 1961)

Vitamin Requirements

The only vitamin which has been considered in relation to environment is vitamin A. Tests with growing chicks by Ascarelli and Bartov (1963) did not show any response to increasing daily intakes of vitamin A at T_A 28–34°C compared with T_A 21–23°C; there was no evidence of any increase or decrease in requirements.

Mention should be made here of ascorbic acid (vitamin C) since it is widely believed that there is a requirement for a dietary supply of this vitamin in warm climates in order to maintain production and prevent thinning of the egg shell. The evidence, reviewed by Kechik and Sykes (1974), is very conflicting; there are some trials in which there has been a well marked response; in others there has been little or no response. There is no obvious feature common to all the positive trials and therefore it is difficult to reproduce the response. It is possible that a positive response occurs only when a level of stress develops to which the particular stock is susceptible.

There is evidence that ascorbate metabolism is affected by some forms of stress, e.g. coccidiosis and starvation (Caudwell and Sykes, 1975), and the levels in some tissues are reduced but it does not follow that ambient temperature within the range normally well tolerated should interfere with synthesis or utilisation in such a way as to create a nutritional requirement. It may be that dietary ascorbate is required only when climatic conditions are such as to give rise to a stress situation but it remains doubtful whether ascorbate should be considered as a nutrient.

Water Requirements

No estimate of the requirement for water has yet been made in that there are no response curves to different daily intakes. It is assumed that voluntary intake satisfies requirements and hence equals these requirements, which may well be correct but it is not necessarily so especially at higher temperatures; voluntary dehydration, a failure to drink enough from freely available supplies, is seen in man in the tropics and could occur in poultry. Water intake under temperate conditions varies widely (ARC, 1975) which is probably as much a reflection of the methods of measurement as of biological variation. The only values quoted in relation to T_A (Wilson, McNally and Ota, 1957) do not, contrary to expectation, show any trends. There are, however, a number of additional values available (Van Kampen and Romijn, 1970; Ito *et al.*, 1970; Hill, 1975; Payne, 1966) and these have been grouped together at 5°C intervals to show average trends (*Figure 2.7*). Up to 25°C the the mean intake per bird is 198 g per day; above that temperature there is an increase of the order of 10 g per d per °C rise. Because of the large variation in measured intake, estimates of water loss were examined as a guide to requirements. The basis for this was data on evaporative water loss (Richards, 1975) made by the gravimetric method under accurately controlled T_A (*Figure 2.7*). Water loss from urine and faeces was calculated (Dixon, 1958) and on the assumption of 50% egg production an addition was made for water lost in the egg. The total non-evaporative water loss amounted to 120 g per day and it was assumed that this remained constant at all T_A. Combining the determined and the calculated components gave an estimate of total water loss (*Figure 2.7*) which is in reasonable agreement with the actual water intake: a fairly constant value (160 g per day) up to 25°C and then a sharp increase, again at the rate of about 10 g per day per °C.

There is obviously a need for more information if water requirements are to be specified but is there any urgency to obtain this information?

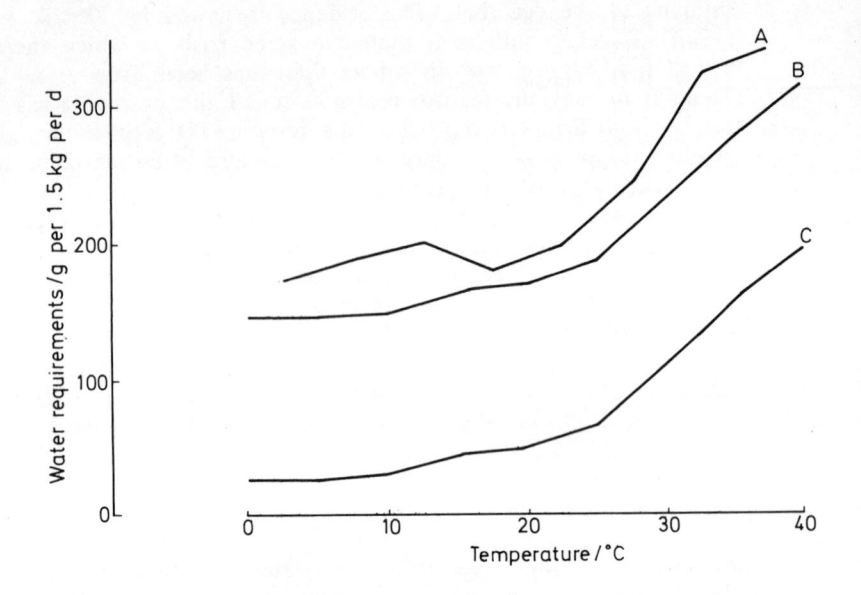

Figure 2.7 Water intake and water loss in relation to ambient temperature. A = mean water intake from 4 trials at 5°C intervals; B = estimated total water loss; C = evaporative water loss (Richards, 1975)

Are there situations in which production is limited by a failure to meet water requirements? Apart from mechanical failure in supply, it would seem unlikely except possibly following a sudden increase in T_A when intake has not adjusted and there is a reduction in food intake and egg production as an immediate consequence. It is possible that temporary water deficits of this type are the means whereby food intake adapts to warm climates.

It is clear, therefore, that much remains to be done before we have an adequate description of the effects of ambient temperature on the nutrient requirements of the laying hen.

References

Agricultural Research Council (1975). *The Nutritional Requirements of Farm Livestock. No. 1. Poultry.* London; Agricultural Research Council
Ascarelli, I. and Bartov, I. (1963). *Poult. Sci.,* **42**, 232
Barott, H.G. and Pringle, E.M. (1946). *J. Nutr.,* **31**, 35
Bray, D.J. and Gesell, J.A. (1961). *Poult. Sci.,* **40**, 1328
Campos, A.C., Wilcox, F.H. and Shaffner, C.S. (1962). *Poult. Sci.,* **41**, 856
Caudwell, F.B. and Sykes, A.H. (1975). Unpublished.
Combs, G.F. (1962). In *Nutrition of Pigs and Poultry,* pp. 141–143. Ed. by J.T. Morgan and D. Lewis. London; Butterworths

Davis, R.H., Hassan, O.E.M. and Sykes, A.H. (1972). *J. agric. Sci., Camb.*, **79**, 363
Davis, R.H., Hassan, O.E.M. and Sykes, A.H. (1973). *J. agric. Sci., Camb.*, **81**, 173
Dixon, J.M. (1958). *Poult. Sci.*, **37**, 410
Emmans, G.C. (1974). In *Energy Requirements of Poultry*, pp. 79–90. Ed. by T.R. Morris and B.M. Freeman. Edinburgh; Longmans
Emmans, G.C., Charles, D.R. and Dun, P. (1975). *Gleadthorpe EHF Poultry Booklet*. Ministry of Agriculture, Fisheries & Food, London
Fisher, C. (1976). *Roche Nutrition Events*. London; Roche Products Ltd London.
Hill, J.A. (1975). *Gleadthorpe EHF Poultry Booklet*. London; Ministry of Agriculture, Fisheries and Food
Hutchinson, J.C.D. and Sykes, A.H. (1953). *J. agric. Sci.*, **43**, 294
Ito, T., Moriya, T., Yamamoto, S. and Mimura, K. (1970). *J. Fac. Fish. Anim. Husb. Hiroshima Univ.*, **9**, 151
Kechik, I.T. and Sykes, A.H. (1974). *Br. Poult. Sci.*, **15**, 449
McFarlane, W.V. (1975). Personal communication
March, B.E. and Biely, J. (1972). *Poult. Sci.*, **51**, 665
Morris, T.R. (1975). Personal communication
Mowbray, R.M. and Sykes, A.H. (1971). *Br. Poult. Sci.*, **12**, 25
Nasir, M. (1975). Personal communication
Olson, D.W., Sunde, M.L. and Bird, H.R. (1972). *Poult. Sci.*, **51**, 1915
O'Neill, S.J.B., Balnave, D. and Jackson, N. (1971). *J. agric. Sci., Camb.*, **77**, 293
Ota, H. (1960). *USDA Publication No. 728*
Payne, C.G. (1966). *Report of the Rural Electrification Conference*, pp. 23–30. London; Electricity Council
Payne, C.G. (1967). In *Environmental Control of Poultry Production*, pp. 40–54. Ed. by T.C. Carter. London; Longmans
Reid, B.L. and Weber, C.W. (1973). *Poult. Sci.*, **52**, 1335
Richards, S.A. (1975). Personal communication
Romijn, C. and Lokhorst, W. (1961). *Tijdschr. Diergeneesk.*, **86**, 153
Romijn, C. and Vreugdenhil, E.L. (1969). *Neth. J. vet. Sci.*, **2**, 32
Shannon, D.W.F. and Brown, W.O. (1969). *Br. Poult. Sci.*, **10**, 13
Smith, A.J. (1973). *Trop. Anim. Hlth. Prod.*, **5**, 259
Swain, S. and Farrell, D.J. (1975). *Poult. Sci.*, **54**, 513
van Es, A.J.H., van Aggelen, D., Nijkamp, H.J., Vogt, J.E. and Scheele, C.W. (1973). *Z. Tierphysiol., Tierernahrg. u. Futtermittelkde.*, **32**, 121
van Kampen, M. and Romijn, C. (1970). *5th Symposium on Energy Metabolism of Farm Animals*, pp. 213–216. EAAP Publication No. 13. Zurich; Juris Druck
Waring, J.J. and Brown, W.O. (1967). *J. agric. Sci., Camb.*, **68**, 149
Wilson, W.O., McNally, E.H. and Ota, H. (1957). *Poult. Sci.*, **36**, 1254

3

CLIMATIC ENVIRONMENT AND POULTRY FEEDING IN PRACTICE

G.C. EMMANS
East of Scotland College of Agriculture, Edinburgh

and

D.R. CHARLES
ADAS, East Midland Region, Derby

Introduction

Feed cost now accounts for about 70% of the cost of production of eggs. Payne (1967) and Smith and Oliver (1972 a and b) found that the feed intake of *ad lib.* fed birds is substantially reduced at high environmental temperatures. Yet Payne (1967) and Mowbray and Sykes (1971) found that up to about 30°C this reduction of feed intake was not associated with a depression of egg production if the intake of the nutrients other than those needed for energy was maintained. Therefore, there has rightly been considerable commercial and experimental interest in the control of laying house temperature in recent years (Marsden *et al.*, 1973; Marsden and Morris, 1975; Spencer, 1975; Clark *et al.*, 1975).

In the UK poultry industry laying house temperature control is always achieved by means of thermostatic variation of the ventilation rate. A minimum air supply intended to be adequate for gas exchange is provided in cold weather, and as house temperature rises the thermostatic control equipment calls for more air, up to a maximum which is considered to be adequate for the removal of metabolic heat in hot weather. This maximum rate is readily calculated from engineering heat balance equations (e.g. Longhouse, Ota and Ashby, 1960; Payne, 1961) and has long been undisputed and successfully applied in practice. However, the minimum rate has been the object of much more contention. Its clarification is extremely important because small changes in winter ventilation rate result in large differences in house temperature and therefore feeding cost.

The series of experiments reported below was intended to provide practical recommendations on the more important aspects of the temperature/nutrition/ventilation interactions. Large groups of birds were used in experiments designed to simulate closely commercial conditions whilst also exploring the underlying biological generalisations.

The first experiment of the series involved a detailed examination of the interaction between diet and climate, and also compared two minimum ventilation rates (one above and one below the commercial rate prevalent at the time of designing the experiment). The second experiment was intended to define the biological and economic optimum temperatures. The third experiment investigated the effects of temperature on birds restricted to less than *ad lib.* feed intake, following the work of Sykes (1972) and considerable commercial interest in restricted feeding. Since both temperature and restricted feeding affect energy balance, it was considered worth exploring the possibility of interaction between them.

For reasons of space the report below deals with only some of the parameters examined. However, the conclusions at the end of this paper are intended to provide some fairly definitive recommendations about house temperature, its control through ventilation, and its implications for feed formulation.

Methods

Eight rooms each containing 1026 birds in cages were separately temperature- and ventilation-controlled.. Ventilation rate was thermostatically varied from minimum to maximum as in commercial houses. In all rooms the maximum ventilation rate was 3 m^3 s^{-1} per 1000 birds (6 ft^3 min^{-1} per bird) and in experiments 2 and 3 the minimum was 0.3 m^3 s^{-1} per 1000 birds (0.75 ft^3 min^{-1} per bird). In Experiment 1, minima of 0.25 and 0.50 m^3 s^{-1} per 1000 birds (0.5 and 1.0 ft^3 min^{-1} per bird) were compared. Experimental temperature treatments were defined as minimum temperatures since heating was provided thermostatically when necassary, but cooling was not provided because it is never used commercially in the UK. Room temperatures were permitted to rise above the normal temperatures when outside temperature exceeded the nominal. Thus the temperature at which the treatments were run closely simulated the temperature patterns which occur in commercial houses at a selection of thermostat settings. At nominal temperatures of 21°C and below the heaters were in practice hardly ever used.

White birds were stocked at 5 per 20 cm cage and brown birds at 4 per 20 cm cage in all three experiments.

EXPERIMENT 1 (1971/72)

This was intended to test Payne's hypothesis (that production is not depressed at high temperature provided that nutrient intake is maintained) for two stocks, and also to find the minimum ventilation requirements for layers. Treatments were as follows:

Stock	Shaver 288, Warren SSL
Minimum temperatures (°C)	16, 24
*Mean temperatures (°C)	19, 24
Minimum ventilation rates (m³ s⁻¹ per 1000 birds)	0.25, 0.5
Dietary protein levels (%)	14, 17 (both at 2.75 kcal per g determined metabolisable energy level, 2.7 N-corrected)

Diets used are given in *Table 3.1*.

EXPERIMENT 2 (1972/73)

This was designed to test the response of *ad lib.* fed birds to temperature more completely by using four temperatures, and including an economic appraisal.

Stock	Babcock B300, Warren SSL
Minimum temperatures (°C)	15, 18, 21, 24
Mean temperatures (°C)	18, 20, 22, 24
Dietary protein levels (%)	15, 18 (both at 2.75 kcal per g classical ME 2.7 N-corrected)

Diets used are given in *Table 3.2*.

EXPERIMENT 3 (1973/74)

This experiment investigated the effects of temperature on restricted feeding as applied by the method of limiting time of access to the trough (the method used by Patel and McGinnis (1970) Bougon (1972) and Bougon and Mevel (1972)).

Stock	Shaver 288, Warren SSL	
Minimum temperatures (°C)	12, 18, 21, 24	
Mean temperatures (°C)	16, 20, 22, 25	
Diets	Metabolisable energy level, classical (kcal per g)	Crude Protein (%)
	2.8	15
	2.8	17
	3.0	15.5
	3.0	18

Diets are given in *Table 3.3*.

*Temperature was recorded every two hours by data logger at 2 points per room outside the cages, and at 12 points in one room in which horizontally the temperature range was within 0.5°C. There was 0.9°C difference between tiers.

34 Climatic environment and poultry feeding in practice

Feeding regimes were *ad lib.*, 8 hours per day access to the trough, and 6 hours per day access to the trough. The access periods were from 08.00 hours to 16.00 and 10.00 hours to 16.00. Lights went off at 17.30. The feeding troughs were covered with wooden lids except for the periods when access was allowed.

All birds in all three experiments were reared on a 10 hour day length and from 20 weeks of age day length was increased by 20 minutes per week up to 17 hours. Light intensity was not less than 5 lux at any point in the room for each experiment.

Energy partitions were calculated in order to find total body heat loss, which was the balancing term in the relationship:

Metabolisable energy intake = egg energy + bodyweight change energy + total heat loss

Table 3.1 Diets used in Experiment 1

Ingredients	%	%
Maize	13.7	11.2
Wheat	50.0	50.0
Barley	10.0	20.6
Ext. soya bean meal	6.2	3.1
White fish meal	6.2	5.0
Meat and bone meal	6.2	2.5
Limestone granules	6.2	6.2
Mineral and vitamin supplement	1.2	1.2
Calculated Analysis		
Metabolisable energy (kcal per g)	2.75	2.75
Crude protein	17.0	14.0
Calcium	3.4	3.1
Phosphorus	0.8	0.6
Fibre	2.0	2.2
Methionine	0.31	0.25
Cystine	0.29	0.26
Lysine	0.84	0.63
Determined analysis		
Crude protein	17.3	14.4
Calcium	3.8	3.4
Phosphorus	0.79	0.6
Crude fibre	3.4	3.3

% Crude protein	Determined % crude protein		Metabolisable energy (kcal per g, N-corrected)
	28–47 weeks	20–60 weeks	
14	15.0 (6)*	14.8 (14)*	2.72 ± 0.014 (7)*
17	17.4 (6)*	17.9 (14)*	2.69 ± 0.019 (7)*

*Number of observations

Table 3.2 Diets used in Experiment 2

Ingredients	%	%
Maize	55.0	55.0
Wheat	15.6	7.1
Hycal	4.5	5.0
Ext. soya bean meal	6.1	9.1
White fish meal	5.0	8.0
Meat and bone meal	5.0	8.0
Limestone granules	7.5	5.9
Minerals and vitamin supplement	1.2	1.8
Calculated Analysis		
Crude protein	15.0	19.0
Calcium	4.0	4.0
Phosphorus	0.6	0.8
Fibre	1.8	1.8
Methionine	0.33	0.44
Cystine	0.21	0.25
Lysine	0.71	0.98
Metabolisable energy (kcal per g)	2.9	2.9
Determined analysis		
Crude protein	14.8	18.2
Calcium	3.95	3.89
Methionine	0.32	0.39
Cystine	0.22	0.28
Lysine	0.78	1.02

Age (weeks)	% Crude protein	Determined % crude protein	Metabolisable energy (kcal per g, N-corrected)
28–43	15	14.6	2.623 (7)*
	19	18.1	2.664 (7)*
44–59	15	14.5	2.683 (8)*
	19	18.3	2.679 (7)*

*Number of observations

Using the energy contents of eggs (1.6 kcal per g) and of bodyweight gain (4 kcal per g) given by Emmans (1974), then:

Heat loss (kcal per bird per day) = ME intake (N-corrected determined value) − [egg ouput (g per bird per day) × 1.6] − [bodyweight gain (g per bird per day) × 4]

Maintenance requirement was calculated assuming an efficiency of 80% for converting dietary metabolisable energy to egg and carcass energy (Emmans, 1974). Thus:

Maintenance requirement (kcal per bird per day) = ME intake − [egg output × 2] − [bodyweight change (g) × 5]

Climatic environment and poultry feeding in practice

Table 3.3 Diets used in Experiment 3

Ingredients	%	%	%	%
Maize	11.2	11.0	55.0	55.0
Wheat	17.1	12.6	17.3	11.8
Barley	47.75	47.9	–	–
Soya bean oil	1.0	1.0	2.0	2.0
Ext. soya bean meal	6.2	7.25	7.0	8.2
White fish meal	5.0	6.2	5.0	7.3
Meat and bone meal	2.5	5.8	5.0	8.0
Limestone granules	8.0	6.75	7.5	5.9
Min. and vit. mix	1.25	1.5	1.25	1.8
600 g Methionine per ton				
Calculated analysis				
ME kcal per g	2.6	2.6	3.0	3.0
CP	14.5	17.0	15.5	18.5
Calcium	3.97	3.91	3.95	3.9
Phosphorus	0.64	0.82	0.70	0.91
Fibre	3.1	3.0	1.9	1.8
Methionine	0.30	0.36	0.33	0.42
Cystine	0.24	0.27	0.22	0.25
Lysine	0.70	0.86	0.74	0.94
Tryptophane	0.16	0.18	0.15	0.17
Determined analysis		*20–36 weeks of age*		
Crude protein	14.9	17.0	15.1	17.8
Calcium	3.78	3.76	3.99	3.89
		36–56 weeks of age		
Crude protein	15.4	17.4	15.8	18.3
Calcium	3.73	3.63	4.03	3.81
		56–72 weeks of age		
Crude protein	15.3	17.3	15.5	17.9
Calcium	3.75	3.89	3.92	3.94
		20–72 weeks of age		
Crude protein	15.2	17.2	15.5	18.0
Calcium	3.75	3.77	3.98	3.89
Dry matter	89.3	89.0	89.0	88.9
Ether extract	3.16	3.30	4.84	5.3
Total ash	12.0	11.3	11.2	11.7
Available carbohydrates	44.8	43.5	45.6	43.6
Metabolisable energy (kcal per g, N-corrected)	2.5	2.6	3.0	2.8

An adjustment was made to the estimate of N-corrected ME intake used in the above relationships to allow for the fact that it was determined on sample cages not the whole flock. Values were corrected back to the N-balance of the flock since the ME determination was for the N-balance of the sample birds.

Table 3.4

Stock		Shaver 288				Warren SSL				Standard error
% Crude protein		14		17		14		17		
Minimum ventilation rate (m³ s⁻¹ per 1000 birds)		0.25	0.5	0.25	0.5	0.25	0.5	0.25	0.5	
Eggs per hen housed	19°C	271	264	270	273	276	272	283	280	
	24°C	253	256	262	265	213	235	216	223	±6.57
Eggs per hen per day	19°C	0.738	0.731	0.745	0.762	0.720	0.711	0.732	0.732	
	24°C	0.716	0.713	0.725	0.738	0.582	0.627	0.579	0.603	±0.0107
Feed intake (g per bird per day)	19°C	108	110	109	111	116	116	117	117	
	24°C	96	96	96	96	98	98	99	100	±1.27
Egg weight (g per egg)	19°C	57.8	58.1	58.9	58.9	58.5	58.5	58.6	59.1	
	24°C	53.6	74.7	54.3	55.0	52.4	53.5	52.8	53.1	±0.45
Mortality (%)	19°C	13.3	15.6	13.3	14.0	5.3	5.0	5.3	6.1	
	24°C	19.5	13.2	14.7	14.7	14.0	10.1	10.7	13.6	±2.75

		Shaver 288		Warren SSL		Standard error
		(means for both ventilation rates)				
Heat loss (kcal per bird) 28–48 weeks of age	19°C	223.1	224.5	242.3	246.9	
	24°C	197.5	192.6	211.2	217.8	±6.91
56–76 weeks	19°C	243.9	241.6	251.5	253.3	
	24°C	209.6	206.0	221.0	215.6	±7.33
Heat loss (kcal per kg) 28–48 weeks	19°C	134.5	129.1	107.0	107.8	
	24°C	126.9	116.9	108.7	107.9	±5.85
56–76 weeks	19°C	134.3	127.0	101.8	103.0	
	24°C	122.3	113.2	101.4	95.7	±8.10
Maintenance requirement (kcal per bird) 28–48 weeks	19°C	205.9	205.8	224.0	227.8	
	24°C	182.4	175.5	197.0	203.3	±7.24
56–76 weeks	19°C	226.8	224.3	237.2	239.7	
	24°C	194.1	191.2	211.4	206.0	±7.67
Maintenance requirement (kcal per kg) 28–48 weeks	19°C	124.2	118.3	98.9	99.5	
	24°C	117.2	106.6	101.4	100.7	±5.76
56–76 weeks	19°C	124.9	117.9	96.0	97.5	
	24°C	113.3	105.1	97.0	91.5	±7.95

At the end of the trials samples of birds were weighed and, after killing, the feathers, combs and wattles removed for separate weighing. Data can be supplied by the authors.

Results and Discussion

Each experiment is dealt with separately in this section but in the Conclusions some general principles have been derived from all three experiments.

EXPERIMENT 1 (20–76 WEEKS OF AGE)

At the lower temperature there was no effect of ventilation rate on any performance trait (*Table 3.4*). In some sub-treatments production was depressed by the low ventilation rate at the high temperature, notably with Shavers on the high protein diet. This may have been due to ammonia levels in the cages, since birds on the high protein diet would have excreted more nitrogen. Manure removal was two or three times per week.

For the Shaver 288, Payne's hypothesis concerning the interaction between temperature and nutrition was confirmed since the effect of temperature on egg production appeared to be through its effect on nutrient intake. This is demonstrated by *Figure 3.1*. Egg production on the high protein diet at high temperature was similar to that on the low protein diet at low temperature, these two treatments having similar protein intakes.

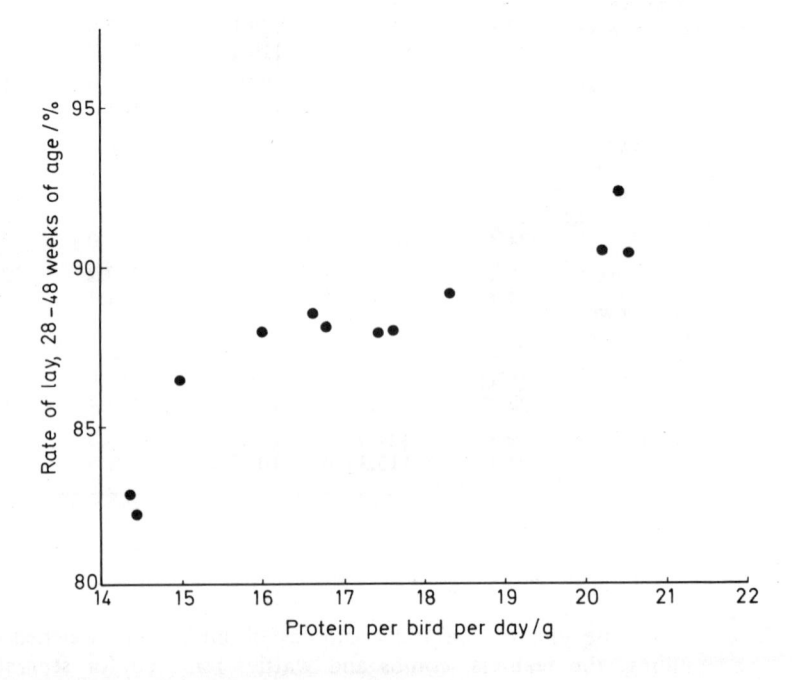

Figure 3.1. Effect of protein intake on rate of lay

Table 3.5

Stock % Crude protein	Babcock B300 18	 15	Warren SSL 18	 15	Standard error
Mean temperature (°C)		*Eggs per hen housed*			
18	251	245	252	244	
20	253	251	249	247	
22	256	247	259	247	
24	250	249	251	243	± 6.95
		Eggs per hen per day			
18	0.748	0.726	0.725	0.705	
20	0.750	0.752	0.713	0.704	
22	0.760	0.736	0.726	0.709	
24	0.742	0.732	0.706	0.692	± 0.0127
		Feed intake per bird per day (g)			
18	111	111	120	122	
20	110	111	117	119	
22	105	107	114	116	
24	99	101	107	109	± 1.85
		Mean egg weight (g)			
18	59.7	59.0	62.6	61.9	
20	59.7	58.4	62.0	61.6	
22	59.0	58.5	61.9	60.7	
24	57.9	57.6	60.1	59.5	± 0.51
		Mortality (%)			
18	19.6	19.6	10.1	11.2	
20	18.4	19.8	8.8	8.8	
22	17.5	19.3	5.0	10.1	
24	14.2	14.9	5.0	7.5	± 3.78
		Egg output (g per bird per day)			
18	44.7	42.8	45.4	43.6	
20	44.8	43.9	44.2	43.0	
22	44.8	43.1	44.9	43.0	
24	43.0	42.2	42.4	41.2	
		*Energetic efficiency** (cal egg output per cal ME intake)			
	A/B	*A/B*	*A/B*	*A/B*	
18	0.242	0.231	0.227	0.216	
20	0.245	0.238	0.227	0.220	
22	0.257	0.242	0.238	0.223	
24	0.261	0.252	0.238	0.227	

*Calculation of energetic efficiency
 grams egg output x 1.6 kcal per egg per bird per day = A
 grams food intake x determined ME (N-corrected) (kcal per g) = B

In the Warren, production was depressed at 24°C, whatever diet was fed. The difference between strains in response to temperature may reflect differences in insulation. The Shavers had less weight of feathers and more weight of comb and wattles per unit of bodyweight than the Warrens (Emmans and Dun, 1973).

At the higher temperature heat loss and maintenance requirement per bird were lower, but in some cases they were similar per kg, as if bodyweight change had been thermoregulatory. The birds on the high protein diet had less heat loss per kg presumably because they were better feathered. (Feathers were weighed at the end of the trial and data can be supplied). Heat losses per bird were generally higher at the end of lay but this was not true per kg of bodyweight.

EXPERIMENT 2

Hen housed egg production was maximal in both stocks at 22°C on the high protein diet and at 20°C on the low protein diet (*Table 3.5*). Hen day rate of lay in Warrens was not much affected by temperature except for a depression at 24°C compared with the other temperatures. In Babcocks, rate of lay was highest at 22°C on the high protein diet and 20°C on the low (*Table 3.5*). Presumably these optimum temperatures were specific to the intakes of the first limiting nutrients with these particular diets.

Feed intake decreased with increasing temperature but not linearly. The rate of decrease was faster the higher the temperature. Over the temperature range studied the decrease in feed intake averaged approximately 1.5% per °C temperature rise (*Table 3.5*).

There was a progressive decrease in egg weight with increase in temperature. At least part of this decrease could probably be attributed to decreased nutrient intake reflected in the reduced feed intake.

Mortality was progressively reduced as temperature was increased. In both stocks birds fed the higher protein diet grew at a faster rate.

There were some cases of depression of shell density due to temperature, some of, but not all of, which probably reflected calcium intake. The percentage of cracked eggs as expressed by packing station seconds was not affected by temperature. In this experiment all treatments received at least 4 g calcium per bird per day.

Four egg price and four feed price markets were arbitrarily chosen and the margin of egg value less feed cost was calculated for each treatment. The egg and feed prices used which were typical at the time of completion of the experiment (early 1974) are given in *Tables 3.6* and *3.7* with their associated margins. Note that this economic appraisal does not take into account seconds. Egg weights were converted to egg grades from the experimental data using the standard deviation method of Wills (1973).

Table 3.6 Babcock B300, Margin of egg income over food cost from 20 to 72 weeks of age (p per bird)

Mean temperature (°C)			18		20		22		24	
% Crude protein	18	15	18	15	18	15	18	15	18	15
Egg market, pence per dozen	Food market, cost per ton (£)		Margin of egg income over food cost (p)							
22	92	80	165	194	170	207	191	207	192	221
	86	80	187	194	192	207	212	207	212	221
	82	70	202	231	207	243	226	242	225	255
	76	70	224	231	229	243	246	242	245	255
28	92	80	302	327	308	343	330	340	327	355
	86	80	324	327	330	343	350	340	347	355
	82	70	339	364	345	379	364	376	360	389
	76	70	361	364	367	379	385	376	380	389

Table 3.7 Warren Sex-Sal-Link, margin of egg income over food cost from 20 to 72 weeks of age (p per bird)

Mean temperature (°C)			18		20		22		24	
% Crude protein	18	15	18	15	18	15	18	15	18	15
Egg market, pence per dozen	Food market, cost per ton (£)		Margin of egg income over food cost (p)							
24	92	80	196	224	198	233	222	241	219	244
	86	80	221	224	222	233	246	241	241	244
	82	70	237	266	238	274	262	280	256	282
	76	70	262	266	262	274	286	280	279	282
30	92	80	340	363	340	373	369	380	360	380
	86	80	364	363	364	373	393	380	382	380
	82	70	381	404	380	414	409	420	397	418
	76	70	405	404	404	414	433	420	420	418

EXPERIMENT 3 (20–72 WEEKS OF AGE)

The third experiment had a complex incomplete factorial design containing so many treatments that it is impossible to report the results fully within the scope of this presentation. Therefore only the important effects and interactions have been selected for detailed reporting. (Complete tables of data can be obtained from the authors).

There was generally no significant depression of hen housed production due to restriction (*Table 3.8*) although at the temperatures associated with the highest levels of production on *ad lib.* feeding there was a non-significant depression of production due to restriction. This was probably not an effect of protein intake as judged by comparison of protein ibtakes with those of *ad lib.* treatments on the low protein diets. The interaction was not significant, suggesting that temperature did not modify the effects of restriction of feed.

42 Climatic environment and poultry feeding in pratice

Table 3.8

	Shaver 288								Standard error
Energy level (kcal per g)	2.8				3.0				
Feeding regime	Ad lib.	Ad lib.	8h	6h	Ad lib.	Ad lib.	8h	6h	
% protein	15	17	17	17	15.5	18	18	18	
°C	Eggs per hen housed								
16	265.5	275.3	275.2	269.9	249.1	261.0	263.2	265.1	
20	266.6	278.6	265.4	259.5	243.4	275.4	257.6	255.7	
22	251.2	268.4	271.1	267.5	241.3	274.9	257.4	257.2	
25	253.2	268.6	261.4	267.4	230.2	249.6	265.9	250.6	±10.91
	Warren SSL								
Energy level (kcal per g)	2.8				3.0				
Feeding regime	Ad lib.	Ad lib.	8h	6h	Ad lib.	Ad lib.	8h	6h	
% protein	15	17	17	17	17.5	18	18	18	
°C	Eggs per hen housed								
16	266.3	266.8	270.1	269.5	258.4	263.3	265.4	257.3	
20	274.6	265.2	269.9	271.8	265.9	262.9	260.7	261.8	
22	254.7	273.1	270.1	268.7	252.4	270.7	259.8	262.6	
25	264.2	265.7	262.8	262.8	253.2	265.4	268.3	267.6	±10.19

In the Shaver 288, production averaged over all feeding treatments fell as temperature was increased (*Table 3.9*). Presumably this was because feed intake was so low that some nutrient became limiting (*Table 3.11*). In the Warren, temperature had no effect on production averaged over all feeding treatments (*Table 3.9*).

Table 3.9

	16°C	20°C	22°C	25°C	Standard error
	Eggs per hen housed				
Shaver 228	265.5	262.8	261.1	255.9	± 3.38
Warren SSL	264.6	266.6	264.1	263.7	± 4.75
	Eggs per hen per day				
Shaver 288	0.753	0.747	0.749	0.783	± 0.0079
Warren SSL	0.741	0.744	0.743	0.736	± 0.0079

The restriction progressively depressed production averaged over all temperatures but not significantly (*Table 3.10*).

Table 3.10

Feeding regime Protein level		Ad lib. Low	Ad lib. High	8h High	6h High	Standard error
	Energy level (kcal per g)	\multicolumn{4}{l	}{Eggs per hen housed}	
Shaver	2.8	259.1	272.7	268.3	266.1	
	3.0	241.5	265.2	261.0	257.1	±10.91
Warren	2.8	264.9	267.7	268.2	268.5	
	3.0	257.5	265.6	263.6	262.3	±10.19
		\multicolumn{4}{l	}{Eggs per hen per day}	
Shaver	2.8	0.739	0.773	0.763	0.742	
	3.0	0.713	0.759	0.748	0.738	±0.0215
Warren	2.8	0.736	0.753	0.742	0.747	
	3.0	0.736	0.746	0.739	0.731	±0.0219

Feed intake fell by 1.3 and 1.0% per °C for Shavers and Warrens respectively in both cases with no interaction between temperature and feed restriction (*Tables 3.11* and *3.12*).

Table 3.11 Feed intake (g per bird per day)

	16°C	20°C	22°C	25°C	Standard error
Shaver	99.8	96.8	93.2	88.5	± 0.47
Warren	123.3	118.4	115.9	111.7	± 0.62

Restriction depressed feed intake in the Shaver significantly by 3% on 8h access to the feed and by 5% on 6h. For the Warren corresponding figures were 4% and 6% (*Table 3.12*). Note that these are based on the cumulative feed intakes from 20 to 72 weeks of age but the restriction was only applied from 40 weeks; thus the percentage reductions in intake were slightly higher during the period when the restriction was applied.

Table 3.12 Feed intake (g per bird per day)

Feeding regime Protein level		Ad lib. Low	Ad lib. High	8h High	6h High	Standard error
	Energy level (kcal per g)	\multicolumn{4}{l	}{Feed intake (g per bird per day)}	
Shaver	2.8	96.6	99.7	96.3	94.4	
	3.0	92.5	94.9	92.3	90.0	± 2.62
Warren	2.8	122.5	122.3	117.2	115.8	
	3.0	117.5	118.7	113.5	111.1	± 2.26

Energy intake of the shaver increased by 6.4 kcal per bird per day for a 100 kcal per kg increase in dietary energy level. For the Warren the corresponding value was 8.7. Both are higher than the increases expected from literature (e.g. Morris, 1968).

Restriction generally had no effect on egg weight and there was no interaction between temperature and restriction. Egg weight was depressed as temperature was increased but it was increased with increasing dietary energy level (*Table 3.13*).

Table 3.13

		16°C	20°C	22°C	25°C	Standard error
		\multicolumn{4}{c}{Egg weight (g)}	
Shaver		57.4	56.8	56.3	54.9	± 0.32
Warren		61.1	61.6	60.7	59.9	± 0.60
Feeding regime		Ad lib.	Ad lib.	8h	6h	
Protein level		Low	High	High	High	
	Energy level (kcal per g)					
Shaver	2.8	55.9	56.5	53.8	55.9	
	3.0	56.0	56.8	57.0	56.9	±0.63
Warren	2.8	60.5	60.2	60.9	60.5	
	3.0	60.6	60.9	61.8	61.1	±0.84
		\multicolumn{4}{c}{Mortality (%)}	
Shaver		7.4	9.3	9.2	8.7	±1.05
Warren		4.7	5.0	5.7	4.0	±2.07
Feeding regime		Ad lib.	Ad lib.	8h	6h	
Protein level		Low	High	High	High	
	Energy level (kcal per g)					
Shaver	2.8	8.3	7.2	7.5	3.6	
	3.0	13.9	8.9	10.3	9.4	±4.01
Warren	2.8	4.2	5.2	3.1	4.5	
	3.0	7.3	5.6	5.2	3.8	±3.37

The general conclusion from the production results of Experiment 3 is that the time of access to the trough method resulted in a mild level of intake restriction which could probably be usefully commercially exploited. Production and egg size were not generally affected by it so that margin of egg value less feed cost could be slightly improved by imposing the restriction. In answer to the question the experiment was intended to clarify it would appear that there is little effect of temperature over the range tested on moderate restriction of feed intake by regulating the time of access to the trough.

Supplementary Field Survey — Predicting Feed Intake in Commercial Flocks

Emmans (1974) gave a prediction equation for ME intake, which includes a term for the effect of temperature based partly on the data reported here. Prediction of intake is essential in feed formulation and this equation emphasises the need to take account of temperature. The equation was validated by comparing predicted and measured intake for 49 8-week periods for 20 commercial laying flocks.

The prediction equation was:

$$ME = W(a + bT) + 2E + 5\Delta W$$

where: ME = ME intake kcal per bird per day
  W = mean period bodyweight, kg
  T = mean environmental temperature, °C
  E = egg output, g per bird per day
  ΔW = body weight gain g per bird per day
  a and b have values of 170, 155 and 140 and −2.2, −2.1 and −2.0 respectively for white, tinted and brown egg laying stocks respectively.

For each 8 week period, measurements were made of W, ΔW, T, E and feed intake. Feed intake was predicted from the equations by assuming that all feeds contained 2.7 kcal ME per g.

Actual feed intake averaged 122.9 g per bird per day with a coefficient of variation (cv) of 9.13%. The ratio of actual to predicted feed averaged 1.030 with a cv of 7.05%.

All flocks were scored for feather loss, L, on a scale from 1 to 6. It was expected that the value of $(a + bT)$, i.e. maintenance per kg per day, in the equation given above might depend on feathering. The ratio of actual maintenance to that predicted was regressed on $\log_e (6 - L)$ and the resulting regression equation was:

$$\frac{\text{Actual maintenance}}{\text{Predicted maintenance}} = 1.405 - 0.28959 \log_e (6 - L)$$

This equation had r = 0.641 (*P*<0.01).

Including feather score in the prediction equations reduced the cv of the ratio of actual to predicted feed intake to 5.46%. Sixty per cent of the flocks had actual feed intakes within ±4% of those predicted.

As a rule of thumb maintenance increased by about 9% for each unit increase in feather loss score. The original production equations fitted best to birds with a feather loss score of 2.0. To derive feed intakes for other feather loss scores the adjustment factors given in *Table 3.14* are used.

Table 3.14

Feather loss score	Multiply estimated maintenance by:
1	0.94
2	1.00
3	1.08
4	1.20
5*	1.40

*Beyond range of data

Supplementary Experiment – Physiology of Temperature Effects

It was not the function of these experiments to explore the physiology in detail, but it was observed that rectal temperature increased as air temperature was increased (*Table 3.15*). It is possible that this is relevant to the depression of egg production frequently found at the highest temperatures used.

Table 3.15

Air temperature (°C)	Rectal temperature (°C)	Standard error
16	41.0	
20	41.1	± 0.065
22	41.3	
25	41.5	

Conclusions and Application

Taking into account all three experiments, it is now possible to make some general conclusions and specific recommendations about laying house temperature.

1. Feed intake of *ad lib.* fed birds falls by about 1.5% per °C temperature rise, but not linearly. The effect is greater the higher the temperature.

2. Payne's hypothesis is tenable, namely that the effect of temperature depends upon the diet offered to the birds. The higher the dietary concentration of the first limiting non-energy nutrient, the higher the temperature associated with maximum egg production. This is a reflection of the effect of temperature on feed intake and emphasises the need to describe diets in terms of nutrient intake rather than just percentage composition.

3. If no nutrient is limiting, there is no depression of production up to at least 22° and in some experiments in this series up to 25°. The precise optimum probably depends on factors such as feather cover, comb size and number of birds per cage, as well as diet. Above the critical level of temperature, production is depressed whatever diet is fed and this may reflect a rise in body temperature.

4. On a given diet egg weight falls as temperature rises, but this is at least partly simply a reflection of nutrient intake.

5. Under most UK market conditions the margin of egg value less feed cost is maximised at a minimum house temperature of 21° (mean 22°), but where the difference in price between egg grades is not important, then higher temperatures would yield more money.

6. Energetic efficiency (defined as yield of egg energy per unit of feed metabolisable energy) improves roughly linearly as temperature rises. Thus it is interesting that maximum biological efficiency and maximum economic efficiency do not necessarily occur at the same temperature.

7. In order to exploit the best temperature, low levels of minimum ventilation rate can safely be used in winter, and 0.25 m^3 s^{-1} per 1000 birds appears to be enough. Practical ventilation systems capable of sufficient wind exclusion to permit this are now available (further information can be obtained from the authors). Note that small errors in the precision of ventilation control cause large depressions of house temperature below the economic optimum (*Figure 3.2*).

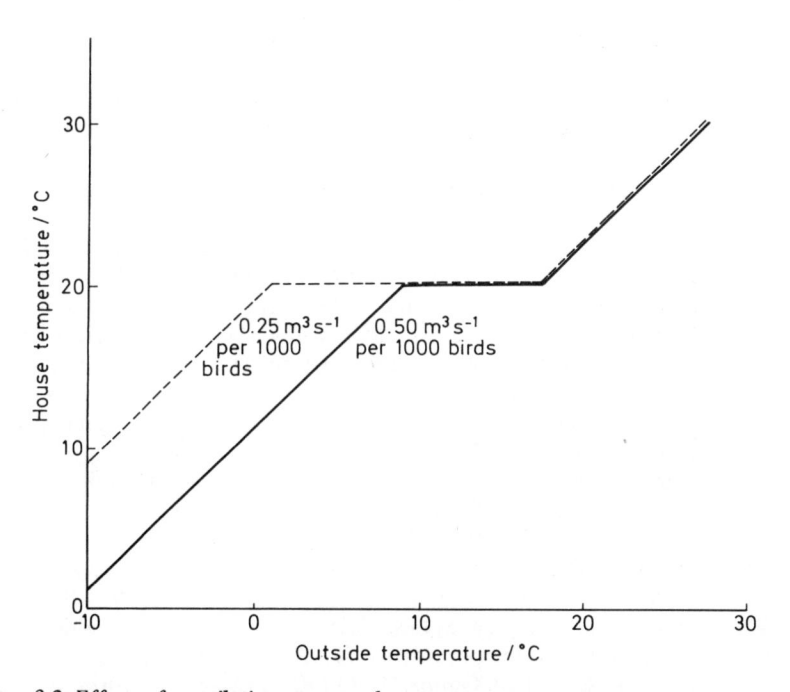

Figure 3.2 Effect of ventilation rate on house temperature

8. There has been widespread uptake of this work by the poultry industry and presumably better and more consistent control of house temperature should permit feed compounders to feed more precisely since feed intake will probably be less variable and more predictable than in the past.

9. Feed intake can be predicted fairly precisely if the standard of feather cover is taken into account as well as the house temperature. Maintenance requirement increases about 9% for each unit increase in feather loss score.

10. To some extent the benefits of temperature and feed restriction can be combined. A mild (less than 5%) restriction can be achieved by limiting time of access to the trough to 6 or 8 hours per day. This feed saving can possibly be exploited without depressing egg production or egg size, but our experiment was not definitive in this respect. There were some non-significant depressions of egg production at the temperatures associated with the highest production for *ad lib.* fed birds.

References

Bougon, M. (1972). 'Influence du rationnement des aliments distribués pendant la periode de production sur les performances des pondeuses.' *Bull. Inf., Ploufragon,* **12** (4), 118

Bougon, M. and Mevel, M. (1972). 'Influence du rationnement en periode de production sur les performances de trois variétés commerciales des pondeuses.' *Bull. Inf., Ploufragon,* **12**, (4), 154

Clark, J.A., Charles, D.R., Wathes, C.M. and Arrow, J. (1975). 'Heat transfer from housed poultry, its implications for environmental control.' *Wld's Poult. Sci. J.,* **31**, 312

Emmans, G.C. (1974). 'The effects of temperature on the performance of laying hens.' In *Energy Requirements of Poultry,* pp. 79–90. Ed. by T.R. Morris and B.M. Freeman. Edinburgh; British Poultry Science Ltd.

Emmans, G.C. and Dun, P. (1973). Temperature and ventilation rate for laying fowls; supplementary report on feathering. Unpublished ADAS data.

Longhouse, A.D., Ota, H. and Ashby, W. (1960). 'Heat and moisture data for poultry housing.' *Agric. Engng St Joseph, Mich.,* **41**, 567

Marsden, A. and Morris, T.R. (1975). 'Comparisons between constant and cycling environmental temperatures applied to laying pullets.' *Wld's Poult. Sci. J.,* **31**, 311

Marsden, A., Wethli, E., Kinread, N. and Morris, T.R. (1973). 'The effect of environmental temperature on feed intake of laying hens.' *Wld's Poult. Sci. J.,* **29**, 286

Morris, T.R. (1968). 'The effect of dietary energy level on the voluntary calorie intake of laying hens.' *Br. Poult. Sci.,* **9**, 285

Mowbray, R.M. and Sykes, A.H. (1971). 'Egg production in warm environmental temperatures.' *Br. Poult. Sci.,* **12**, 25

Patel, P.R. and McGinnis, J. (1970). 'Effect of restricting feeding time on feed consumption, egg production and body weight gain on Leghorn pullets.' *Poult. Sci.,* **49**, 1425

Payne, C.G. (1961). 'Studies on the climate of broiler houses. 2. Comparison of ventilation systems.' *Br. Vet. J.,* **117**, 106

Payne, C.G. (1967). 'Environmental temperature and egg production.' In *The Physiology of the Domestic Fowl,* pp. 235–241. Ed. by C. Horton-Smith and E.C. Amoroso. Edinburgh; Oliver and Boyd

Smith, A.J. and Oliver, J. (1972a). 'Some nutritional problems associated with egg production at high environmental temperatures. 1. The effect of environmental temperature and rationing treatments on the productivity of pullets fed on diets of different energy content.' *Rhod. J. Agric. Res.*, **10**, 3

Smith, A.J. and Oliver, J. (1972b). 'Some nutritional problems associated with egg production at high environmental temperatures. 4. The effect of prolonged exposure to high environmental temperatures on the productivity of pullets fed on high energy diets.' *Rhod. J. Agric. Res.*, **10**, 43

Spencer, P.G. (1975). 'A cost benefit analysis of temperature controls in housing laying hens.' *Wld's Poult. Sci. J.*, **31**, 309

Sykes, A.H. (1972). 'The energy cost of egg production.' In *Egg Formation and Production*, pp. 187–196. Ed. by B.M. Freeman and P.E. Lake. Edinburgh; British Poultry Science Ltd

Wills, J.R. (1973). Personal communication

4

THE INFLUENCE OF CLIMATIC VARIABLES ON ENERGY METABOLISM AND ASSOCIATED ASPECTS OF PRODUCTIVITY IN THE PIG

C.W. HOLMES
Massey University, New Zealand

and

W.H. CLOSE
ARC Institute of Animal Physiology, Cambridge

Introduction

The most important direct effect of climatic factors on the pig is on the exchange of heat between the pig and its environment.

The relation of heat loss to productivity can be summarised by the equation:

Energy retained in body = metabolised energy eaten − heat lost

It follows therefore that climatic factors which influence the heat transfer of the pig will also influence the rate at which energy is retained and the efficiency with which dietary energy is utilised for productive purposes. The aim of this article is to discuss the effects of the various climatic factors on the heat transfer of the pig and the associated effects on several aspects of productivity.

Exchange of Heat Between the Pig and its Environment

Heat exchange is regulated so that over a period of time the heat produced within the body is equal to the amount of heat lost from the body, enabling body temperature to remain relatively constant. This balance is drawn schematically in *Figure 4.1* which shows that heat production and heat loss take place in several ways, and illustrates the role of the temperature-regulating system in maintaining balance.

1. (a) Vital activities
   (b) Heat production with food intake
   (c) Muscular activity
   (d) Productive processes
2. Extra thermoregulatory heat production

   HEAT PRODUCED WITHIN BODY

A. Evaporative losses
   Sweating, panting, wet skin

B. Non-evaporative losses
   Convective, radiative, conductive, heat of warming

   HEAT LOST FROM BODY

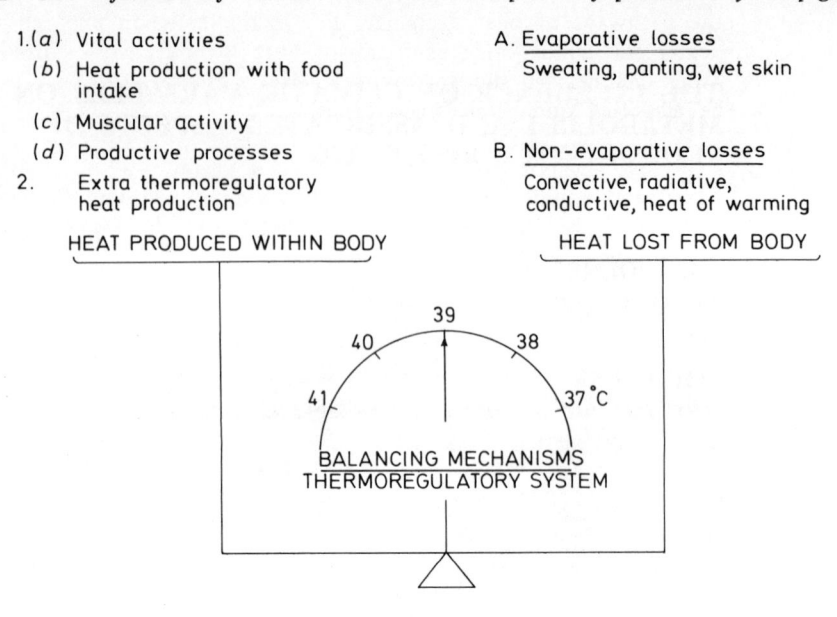

Figure 4.1 The balance of heat production and heat loss in the pig

HEAT PRODUCTION WITHIN THE BODY

Heat is produced within the body as a result of many processes: these include the vital metabolic processes (respiration, circulation, etc.), muscular activity, the ingestion and digestion of food and the utilisation of nutrients for such processes as growth, reproduction and lactation (*Figure 4.1*). In addition, if the animal is kept in a cold environment, it may be forced to produce extra heat in order to maintain its body temperature at a constant level; this is termed *extra thermoregulatory heat production*. It is quite distinct from the other types of heat production because it is produced specifically to meet the environmental demand for heat, whereas in all other cases heat is produced as an inevitable by-product of the metabolic activities within the body. On the other hand, in a hot environment it may become difficult for the pig to dissipate sufficient heat to the environment. In these circumstances body temperature will rise and, as a consequence, heat production may increase thereby further compounding the stressful nature of the environment.

HEAT LOSS FROM THE BODY

Generally the pig's body is at a higher temperature than the environment so that heat is lost by all channels. In certain conditions, however, the animal may actually gain heat from the environment by one or more of the pathways of heat exchange. For example, when exposed to a radiant heater the animal may gain radiant heat from it, while losing heat by the other channels to the environment. These channels of heat exchange can be classified under two main types:

1. Non-evaporative, or sensible heat transfer, by radiation, convection and conduction. The 'heat of warming' required by ingested materials should also be included under this heading.
2. Evaporative heat transfer.

The heat exchanges of the pig have been discussed in detail by Mount (1968), and the relative importance of the channels of heat transfer is illustrated by the data of Bond, Kelly and Heitman (1959) (*Table 4.1*).

Table 4.1 The partition of the total loss of heat from pigs (range of live weight 30–200 kg) into the four major components, at three air temperatures (Bond, Kelly and Heitman, 1959)

Air temperature (°C)	Radiation	Convection	Conduction[1]	Evaporation
4	35	38	13	15
21	27	34	11	28
38	3	5	3	90

[1]Heat of warming may be included in this component

THE BALANCE OF HEAT PRODUCTION WITH HEAT LOSS

The mechanisms which the pig can utilise in order to maintain the balance between heat produced and heat lost fall conveniently into two main classes: those which control the rate of heat production in the body, and those which control the rate of heat loss from the body. The effects of these changes in heat production and heat loss are illustrated in *Figure 4.2* in relation to air temperature. Some features in *Figure 4.2* require definition.

1. The zone of thermoneutrality can be defined as the range of temperature within which heat production is independent of air temperature. Within this zone, heat production occurs at a rate which depends primarily on the level of feeding and the live weight of the pig.
2. The lower critical temperature (T_{cl}) marks the lower end of the thermoneutral range; at temperatures below T_{cl} the pig must increase its rate of heat production to maintain thermal equilibrium. Under these conditions heat production becomes increasingly dependent on air temperature and the rate at which it increases is indicated by the slope of the line relating H (heat production) to temperature, below T_{cl}.

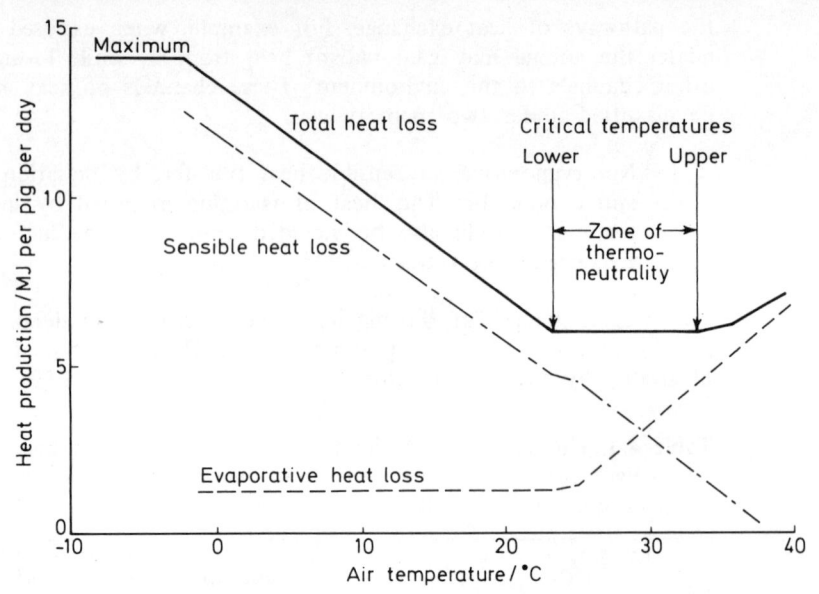

54 *The influence of climatic variables on aspects of productivity in pigs*

Figure 4.2 A diagrammatic representation of the relation between air temperature and the components of heat loss in the pig. For a pig 60 kg live weight with ME intake of 2 × maintenance

3. The maximum rate of heat production of which the pig is capable and the temperature at which this occurs is indicated; if the temperature falls below this level the pig will die of cold stress.
4. The upper critical temperature (T_{cu}) represents the highest temperature at which the pig is able to keep its body temperature reasonably constant; above this level both body temperature and heat production increase.

Definitions concerning thermoneutral environments are likely to vary depending on the context; this has been the subject of a recent discussion (Mount, 1974).

COMBINED EFFECTS OF ANIMAL AND CLIMATIC FACTORS ON THE EXCHANGE OF HEAT

The interactions between the animal and its environment are illustrated in *Figure 4.3* which shows two important points.

1. If the minimal, thermoneutral, level of heat production is increased (for example by an increase in the level of feeding) the values for both the lower and upper critical temperatures are decreased.
2. If, at low temperatures, the rate of increase in heat loss from the body is decreased (for example for a group of pigs compared with an individual pig) the value for the lower critical temperature is

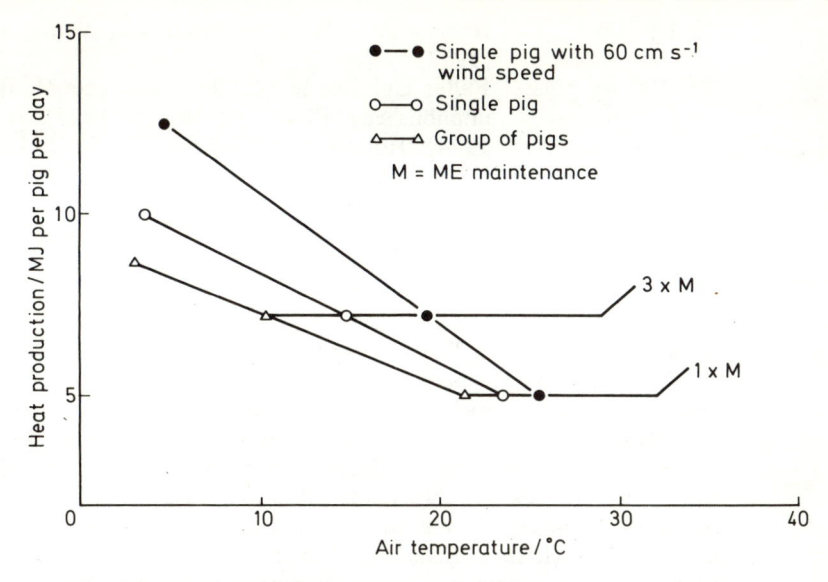

Figure 4.3 A diagrammatic representation of the effects of some animal and climatic factors on the heat production of the pig (60 kg live weight)

decreased. If, on the other hand, this rate is increased (for example by an increase in the rate of air movement) the value for the lower critical temperature is increased. Thus it is apparent that both climatic and animal factors are important in their effects on heat exchange, and these may interact in their effects on the pig's responses to its climatic environment.

Analysis of Experimental Results for Heat Production and Heat Loss of Pigs

It is now proposed to analyse published values for heat production and heat loss to determine the effects of climatic factors on the energy metabolism of pigs.

HEAT PRODUCTION IN THE ZONE OF THERMONEUTRALITY

There is available in the published literature a considerable amount of information about the energy metabolism of pigs ranging from birth to 100 kg live weight, and some information about the energy metabolism of sows. This has been analysed in relation to the live weight and age of pigs, their intake of metabolisable energy, and the effects of pregnancy in sows.

The material has been drawn from the following sources:

Baby pigs: Jordan and Brown (1970); Jordan (1974); Kielanowski and Kotarbinska (1970)

15 kg pigs: Burlacu et al. (1973); Jenkinson, Young and Ashton (1967)
20–100 kg pigs: Fuller and Boyne (1972); Close and Mount (1976, unpublished); Close and Mount (1975); Holmes (1973, 1974); Holmes and Breirem (1973); Verstegen et al. (1973); Kielanowski and Kotarbinska (1970); Thorbek (1974, 1975)
Sows: Holmes and McLean (1974); Verstegen, van Es and Nijkamp (1971)

The values presented in *Table 4.2* represent the mean values for the particular conditions. Those for pregnant sows were calculated by adding to those for non-pregnant sows a value of 1.05 per kg$^{0.75}$ daily per day since conception, which was derived from the combined results of Holmes and McLean (1974) and Verstegen, van Es and Nijkamp (1971). Only values for thermoneutral conditions were used in compiling this table. In some cases it was possible to establish thermoneutrality for the experiment in question; in other cases it was necessary to extrapolate from other experiments. There is little available information about the energy metabolism of pigs in some categories, in particular within the range of live weight 10 to 20 kg, immediately after weaning from milk, and sows either pregnant or in lactation.

Table 4.2 Heat production, under thermoneutral conditions, of pigs at several live weights and fed on different amounts of energy

Type of pig	\multicolumn{4}{c}{Metabolisable energy intake}			
	0	M[1]	2M	3M
	\multicolumn{4}{c}{Heat production (MJ per kg$^{0.75}$ daily)}			
MILK-FED				
Newborn	0.531	0.573	0.640	0.707
Young	0.406	0.494	0.565	0.636
SOLID-FED				
Young		0.649	0.795	0.941
20–50 kg	0.397	0.423	0.561	0.699
50–100 kg	0.364	0.410	0.527	0.644
SOWS				
120–180 kg		0.393	0.531	0.699
PREGNANT SOWS				
60 days		0.456	0.594	0.732
112 days		0.510	0.653	0.787

[1] M = metabolised energy required for maintenance; assumed to be 0.42 MJ ME per kg$^{0.75}$ daily

The values in *Table 4.2* have been presented relative to live weight$^{0.75}$ for comparative purposes; from these the heat production of pigs weighing 20 to 180 kg live weight, fed at maintenance or above and not pregnant or lactating, can be predicted by the equation:

$$H = 0.27 + 0.32 \text{ ME}$$

where H = heat produced, MJ per kg$^{0.75}$ daily
ME = metabolisable energy eaten, MJ per kg$^{0.75}$ daily

For subsequent calculations with pigs of various live weights it is more convenient to calculate values for heat production *per pig* daily; these values, together with the corresponding approximate intakes of ME, are presented in *Table 4.3*.

The values in *Table 4.3* refer to the total heat production of pigs; however in calculations associated with environmental control in pig houses it is necessary to have information about the partition of this total heat production between the evaporative and non-evaporative components. *Table 4.4* has been compiled for this purpose, from the results of Holmes (1966), Close (1970) and Bond, Kelly and Heitman (1952), for different air temperatures. This partition, in particular at higher temperatures, will depend to some extent on the design of the pig house; for example poor drainage in a house will cause more evaporation to occur from urine and water, and therefore a higher percentage of total heat loss to occur as evaporative loss. Nevertheless, the values in *Table 4.4* have been drawn from experiments in which pigs were kept in groups under conditions which were reasonably similar to those encountered in farm buildings. They are therefore applicable, in conjunction with the values in *Table 4.3*, in calculations concerned with environmental control in pig houses.

HEAT LOSS; CALCULATION OF $H \triangle T^{-1}$

The published results of experiments in which the energy metabolism of pigs has been studied in relation to climatic variables have been analysed. Calculations have been performed on the basis of H m^{-2} °CΔT^{-1} subsequently abbreviated to $H \Delta T^{-1}$, where: H = total heat loss or production (MJ per day); m^{2} = surface area of pig calculated as 0.097 W kg$^{0.633}$ (Brody, 1945); and °CΔT = difference between rectal temperature and air temperature (rectal temperature was assumed to be 39°C if not measured).

The values for heat loss should strictly refer only to sensible loss since evaporative loss is not dependent primarily on a temperature difference. However, the values actually used were those for total heat loss on production because in many experiments only total heat loss on production was measured and because the interest in the present context is in total energy exchange.

In the majority of experiments in which the influence of air temperature on heat exchange has been investigated, the air temperature has remained at a constant level during each measurement. This is not

Table 4.3 Heat production, under thermoneutral conditions, of pigs of several live weights and fed on different amounts of energy

Type of pig	Live weight of pig (kg)	Metabolisable energy intake						
		0	M^1		2M		3M	
		Heat production (and approximate intake of metabolised energy) (MJ per pig daily)							
MILK-FED								
Newborn	2	0.89	0.96	(0.7)	1.08	(1.4)	1.19	(2.1)
Young	4	1.15	1.40	(1.2)	1.59	(2.4)	1.80	(3.5)
SOLID-FED								
	15		4.94	(3.2)	6.06	(6.4)	7.18	(9.6)
	20	3.76	4.00	(4.0)	5.31	(7.9)	6.61	(11.9)
	60	7.85	8.84	(9.0)	11.37	(18.0)	13.89	(27.1)
	100	11.51	13.00	(13.2)	16.67	(26.5)	20.37	(39.7)
SOWS	140		16.01	(17.0)	21.63	(34.1)	27.25	(51.1)
PREGNANT SOWS								
60 days	140		18.56		24.18		29.80	
112 days	140		20.77		26.56		32.02	

[1] $M = 0.42$ MJ ME per $kg^{0.75}$ daily

Table 4.4 The partition of total heat loss from a pig pen into its evaporative and non-evaporative components, at several air temperatures (Holmes, 1966; Close, 1970; Bond, Kelly and Heitman, 1952)

Air temperature (°C)	Percentage of total heat loss from pen as:	
	Non-evaporative	Evaporative
5	85	15
10	84	16
15	80	20
20	71	29
25	58	42
30	43	57
35	20	80

normally the situation encountered in practice. Nevertheless Morrison, Heitman and Givens (1975) showed that the effect of a cycling temperature on the rate of gain in live weight of pigs was similar to that of a constant temperature equal to the mean value of the cycle; the validity of this similarity of effects is implicit in the present discussion of heat transfer.

The Effects of Air Temperature on $H \Delta T^{-1}$

Values for $H \Delta T^{-1}$ have been calculated for individual pigs and for groups of pigs of several sizes, and the average results for different temperatures are presented in *Table 4.5*. These values were measured under controlled conditions, with low levels of air movement, mean radiant temperature similar to mean air temperature, no straw or other bedding material used and a variety of floor types.

The material was drawn from:

Baby pigs: Mount (1960, 1963)
Pigs 11–100 kg: Close (1970); Close and Mount (1976, unpublished); Holmes (1966); Verstegen (1971); Verstegen *et al.* (1973)
Sows: Holmes and McLean (1974)

Comparisons made within experiments of values for total heat production (H) with the values for $H \Delta T^{-1}$ indicate that $H \Delta T^{-1}$ continues to decline at temperatures below the lower critical temperature (T_{cl}). Part of this decline is almost certainly due to further changes in the posture of the animals in response to cold conditions, with consequent decreases in the area of surface exposed for transfer of heat. The increase in $H \Delta T^{-1}$ at higher temperatures is associated with an increase in the rate of evaporative heat loss from the pigs.

Table 4.5 Rates of total heat loss[1] per unit surface area from pigs of several live weights, measured at different ambient temperatures, expressed relative to the difference in temperature between the pig and the environment ($H \, \Delta T^{-1}$)

	Air temperature (°C)						
Type of pig	0–4	5–9	10–14	15–19	20–24	25–29	30–34
	$H \Delta T^{-1}$ (kJ m^{-2} °CΔT^{-1} per day)

INDIVIDUAL PIGS

0–7 days		460			544		711
20–50 kg	360	448	456	527	510		925
50–100 kg	360	381	456	527	527		858
Sows: Fat		339	381	444	548		
Thin		423	460	477			

GROUPS OF PIGS

Newborn			293		322		602
20–50 kg		377	435	490	561	602	996
50–100 kg		368	423	490	531	602	885

[1] Measured under controlled climatic conditions: (*a*) low levels of air movement; (*b*) mean radiant temperature similar to air temperature; (*c*) no straw or other bedding material and various types of floor

The effects of climatic and animal factors on $H \, \Delta T^{-1}$

Rate of air movement Mount (1966) showed that for individual newborn piglets at air temperatures of 20°C and 30°C the value for $H \, \Delta T^{-1}$ was increased by 12–16% by wind speeds of 35–82 cm s^{-1}, and by 19–38% by a wind speed of 158 cm s^{-1}. Bond, Heitman and Kelly (1965) showed that at 16°C a wind of 150 cm s^{-1} increased $H \, \Delta T^{-1}$ by 35% for groups of pigs weighing 70–150 kg. However, Holmes (1966) found that a wind of 25 cm s^{-1} had no effect on the heat production of groups of pigs, 30–50 kg live weight, at 12°C.

Bedding and floor type Verstegen and Van der Hel (1974) measured the heat production of groups of young pigs, 30–40 kg live weight, at 10–13°C when housed on either concrete, asphalt or asphalt with straw bedding; the values for $H \, \Delta T^{-1}$ were reduced by 10 and 16%, respectively for the latter floors when compared with concrete. Stephens (1971) showed similar effects of straw bedding for newborn piglets.

These effects are also illustrated by some results from Mount (1967) who measured the rate of heat flow from pig to floor, with different floor materials, at different temperatures beneath the floor. Heat flow increased by 50–100% at 20°C compared with 30°C ambient temperature; heat flow through a wooden floor was about 50% of that through a concrete floor; and through a floor of expanded polystyrene it was only 15–20% of that through a concrete floor. If pigs are forced to lie in wet conditions, conduction through the floor and evaporation will be increased.

Animal factors Age or live weight, body fatness and the grouping of pigs together all exert effects on $H \Delta T^{-1}$ which are illustrated in *Table 4.5*. The effect of fatness on $H \Delta T^{-1}$ suggests that the development of pigs with leaner carcases may result in these animals being less able to tolerate cold environments than the so-called 'unimproved' pig with larger quantities of subcutaneous fat in its carcase.

Radiant environment Mount (1964) showed that for individual baby pigs, a change of 1°C in the mean temperature of surrounding surfaces had a similar effect on heat production to a change of 1°C in air temperature. In an uninsulated building the temperatures of the internal surfaces of the roof, walls and windows are likely to be higher than the internal air temperature in hot, sunny weather, but lower than the internal air temperature in cold weather. In these cases the approximate mean environmental temperature can be calculated as:

$$\text{Mean environmental temperature} = \frac{(3 \times T_{\text{air}}) + T_{\text{surfaces}}}{4}$$

(This calculation assumes that the roof, walls and windows account for 50% of the radiant environment of the pig.) It can be calculated in this way that if air temperature is 20°C and the mean temperature of the surfaces is 10°C, the mean environmental temperature is 17.5°C; if the surfaces are at 30°C the mean environmental temperature is 22.5°C. These effects will reduce the values of air temperature for T_{cl} and the upper critical temperature (T_{cu}) (*Tables 4.8, 4.9, 4.10*) in conditions where internal surface temperatures are higher than the internal air temperature within a building, but will increase values for T_{cl} and T_{cu} in conditions where the internal surfaces are at a lower temperature than the internal air temperature.

Holmes and McLean (1976) found that at 10°C the effect on total heat production of groups of young pigs (11 kg live weight) of a shiny reflective aluminium foil lining to walls and ceilings was equivalent to an increase in air temperature of approximately 2°C.

Stephens and Start (1970) found that exposure of newborn piglets to a radiant lamp heater was equivalent, in terms of heat production, to an increase in air temperature of 14–17°C. However, it has also been shown (Gilbert, cited by Mount, 1968) that in practice use of such lamps is likely to be associated with an increase in the rate of air movement

Table 4.6 Mean estimated values[1] for $H\Delta T^{-1}$ for pigs of several categories: (a) at the lower critical temperature; (b) the minimum values attained at temperatures at least 10°C below the lower critical temperature

Type of pig	$H\Delta T^{-1}$ (kJ m^{-2} °CΔT^{-1} daily)	
	(a)	(b)
INDIVIDUAL PIGS		
Newborn	753	460
20–100 kg	460	356
Sows: Fat	460	335
Thin	502	418
GROUPS OF PIGS		
Newborn	481	293
20–100 kg	418	335

[1] Measured under controlled climatic conditions; see Table 4.5

Table 4.7 Mean estimated values for rates of heat loss[1] per pig ('$H\Delta T^{-1}$ per pig') for pigs of different categories: (a) at the lower critical temperature; (b) at temperatures more than 10°C below the lower critical temperature

Live weight (kg)	'$H\Delta T^{-1}$ per pig' (kg °CΔT^{-1} per pig daily)	
	(a)	(b)
INDIVIDUAL PIGS		
2	113	69
20	300	231
60	594	459
100	828	640
140 Fat	1017	740
140 Thin	1110	925
GROUPS OF PIGS		
2	72	44
20	272	218
60	540	431
100	753	602

[1] Measured under controlled climatic conditions; see Table 4.5

beneath the lamps. Such effects were excluded by the design of the equipment used by Stephens and Start, so that the probable beneficial effect of these lamps is less than measured by them.

Temperature of food and liquid and 'heat of warming' Holmes (1971a) showed that the growth rate of pigs fed whey at 16°C was 11% less than that of pigs fed whey at 40°C, at 16°C air temperature; the difference was 5% for pigs housed at 22°C. However, Forbes and Walker (1968) showed that the temperature (5, 13 or 35°C) at which a mixture of meal and water was fed to pigs had no effect on growth rates. It is likely that this component of the environment will be important only when large volumes of cold liquids are drunk, particularly at low air temperatures.

CALCULATION OF VALUES FOR LOWER CRITICAL TEMPERATURE (T_{cl})

The values for $H\Delta T^{-1}$ in *Table 4.5* were calculated from the results of experiments in which heat production or heat loss was measured at various air temperatures. It was possible therefore in the majority of cases to estimate for each experiment a value for the lower critical temperature (T_{cl}, see *Figure 4.2*). Values for $H\Delta T^{-1}$ at the T_{cl} have been estimated, and these are presented in *Table 4.6* together with estimates of the value for $H\Delta T^{-1}$ at a temperature 10°C below T_{cl}, which are considered here to be the minimal attainable values. The values for $H\Delta T^{-1}$ have been converted into units of heat production *per pig* per day per °CΔT or '$H\Delta T^{-1}$ per pig', and these are presented in *Table 4.7* for the several categories of live weight. By combining these latter values with those for total thermoneutral heat production in *Table 4.3* it is possible to calculate approximate values of T_{cl} for pigs of several live weights and fed on several levels of ME intake according to the equation:

$$T_{cl} = T_R - \left(\frac{\text{Thermoneutral heat production per pig}}{\text{'}H\Delta T^{-1}\text{' per pig' at lower critical temperature}} \right)$$

where T_R = rectal temperature, taken to be 39°C.

The calculated values for T_{cl} are presented in *Table 4.8* for pigs individually or in groups; these values, together with those in *Table 4.7*, characterise the effects of air temperature, level of feeding, and individuals versus groups on heat production. No account has been taken of the possible effect of the number of pigs in the group, which ranged from 3 to 6 in the experiments cited above.

In addition to the effects of live weight, level of ME intake and body fatness on T_{cl}, which are illustrated in *Table 4.8*, this table shows that the value for T_{cl} decreases as a result of pregnancy; there is in fact little experimental evidence for this effect.

Table 4.8 Calculated values[1] of air temperature corresponding to lower critical temperatures (T_{cl}, °C) for individual pigs of different categories, on three different levels of energy intake

	Live weight (kg)	M[2]	2M	3M
		\multicolumn{3}{c}{T_{cl} (°C)}
	2	31 (−4)	29 (−5)	29 (−5)
	20	26 (−2)	21 (−2)	17 (−2)
	60	24 (−1)	20 (−2)	16 (−3)
	100	23 (−1)	19 (−2)	14 (−2)
Pregnant sows	140			
Thin 0 days		25	20	14
60 days		22	17	12
112 days		20	15	11
Fat 0 days		23	18	12
60 days		21	15	10
112 days		19	13	7

Values in parentheses are the effects due to the pigs being in a group of 3 to 6 pigs in each case
[1] Calculated for controlled climatic conditions; see Table 4.5
[2] See Table 4.2

Table 4.9 Assumed maximum rates of heat loss[1] from individual non-hyperthermic pigs at high temperatures and corresponding values for air temperatures corresponding to upper critical temperature[1] at several levels of energy intake

	Live weight (kg)	Maximum rate of heat loss (kJ °CΔT^{-1} per pig daily)	M[2]	2M	3M
			\multicolumn{3}{c}{T_{cu} (°C)}
	2	151	33	32	31
	20	653	33	31	30
	60	1297	32	30	29
	100	1807	32	30	28
Pregnant sows	140				
0 days		2218	32	29	27
112 days		2218	30	27	25

[1] Calculated for controlled climatic conditions; see Table 4.5
[2] See Table 4.2

The combined effects of some climatic factors

The effects of wind speed and of floor type on critical temperature have been calculated assuming that winds of 30–80 cm s^{-1} increase $H\Delta T^{-1}$ by 15% and that straw bedding and asphalt floor reduce $H\Delta T^{-1}$ by 15% (*see* pp. 60 and 61). These calculations indicate that T_{cl} is reduced by straw and asphalt by 1–2°C at lower levels of feeding and by 3–4°C at higher levels; increased air movement had opposite effects of similar magnitude. The values for T_{cl} for 20 kg pigs (*Table 4.8*) are in reasonable agreement with those calculated for 35 kg pigs by Mount (1975) using a rather different approach. Although these values are estimates they provide information which is likely to be useful in practice.

CALCULATION OF VALUES FOR UPPER CRITICAL TEMPERATURE (T_{cu})

To calculate T_{cu} (*see Figure 4.2*) it is necessary to know, as above, for T_{cl}, the minimal level of heat production for the particular pig, and in this case the *maximum* rate at which the pig can lose heat, while not under excessive stress in hot conditions.

The results of Holmes (1966) and Close (1970) for individual pigs and groups of pigs show that at 30°C air temperature, 31–61% RH, the pigs were able to lose heat at 858–1110 kJ m^{-2} °CΔT^{-1} daily. Rectal temperature was increased by 0.3–0.6°C above the level measured at 20°C, respiratory rate increased to between 35 and 105 per min, and neither total heat loss nor growth rate was affected at 30°C. However, Holmes (1973, 1974) showed that exposure of pigs weighing 30–70 kg to air temperatures of 32–34°C caused increases in rectal temperature of 1–2°C and decreases in growth rate and energy retention; the rate at which heat

Table 4.10 The calculated ranges of thermoneutral air temperatures (°C)[1] for pigs in various categories

	Live weight (kg)	Metabolisable energy intake		
		M[2]	2M	3M
		Thermoneutral zone		
	2	31–33	29–32	29–31
	20	26–33	21–31	17–30
	60	24–32	20–30	16–29
	100	23–32	19–30	14–28
Pregnant sows 112 days	140			
	Thin	20–30	15–27	11–25
	Fat	19–30	13–27	7–25

[1] Calculated for controlled climatic conditions; see Table 4.5
[2] See Table 4.2

was dissipated by these pigs was approximately 1300 kJ m^{-2} °CΔT^{-1} daily. It may be concluded that a value of 1000 kJ m^{-2} °CΔT^{-1} daily represents the highest rate of heat loss, or maximum $H\Delta T^{-1}$, which can be maintained by the pig with a dry skin. Values for T_{cu} have been calculated by subtracting from rectal temperature (assumed to be 39°C) the quantity:

$$\frac{\text{Thermoneutral heat production per pig}}{\text{Assumed maximum '}H\Delta T^{-1}\text{ per pig'}}$$

These values together with those for maximum '$H\Delta T^{-1}$ per pig' are presented in *Table 4.9* for pigs of several live weights.

It is important to note that the value of 1000 kJ m^{-2} °CΔT^{-1} daily was derived from measurements with pigs weighing from 20 to 60 kg, and that this value has been extrapolated for use with pigs ranging from 2 to 140 kg (*Table 4.9*). The validity of this extrapolation requires experimental verification. Pigs with wet skin can increase the maximum rate of heat loss considerably due to the increased evaporation from the skin (*see below*); this would increase the value of T_{cu}.

From the values in Tables *4.8* and *4.9*, the thermoneutral zone can be described; these values are presented in *Table 4.10* for individual pigs.

The effect of other climatic factors on upper critical temperature

A climatic factor, e.g. wind, which increases the rate of heat loss from the pig, can be expected to have an adverse effect on the pig which has already been forced to increase its heat production by exposure to a cold environment. However the same factor may be beneficial to the pig which is finding difficulty in dissipating sufficient heat to a hot environment. Because the pig relies to a greater extent on evaporative heat loss at high temperatures, relative humidity becomes important under these conditions. Evidence on this aspect has been provided by Ingram (1965a) and Morrison, Bond and Heitman (1968) which suggests that at 30°C air temperature an increase of 18% in relative humidity (or approximately 3°C wet bulb) is equivalent to an increase of 1°C in air temperature with respect to the effects on heat losses.

A factor which is likely to be of great importance under hot conditions is the degree of wetness of the skin of the pig. Ingram (1965b) showed that under hot conditions the rate of evaporation from the skin of a pig wetted with mud or water was increased to a rate at which all the heat produced by the pig could have been dissipated by evaporation from only a part of the pig's total surface.

THE EFFECTS OF CLIMATIC FACTORS ON SEVERAL ASPECTS OF PRODUCTIVITY

Energy metabolism

From the values for '$H\Delta T^{-1}$ per pig' presented in *Table 4.7* it is possible to calculate the rate at which heat production increases as temperature falls below T_{cl}. These have been calculated (*Table 4.11*) over the range of temperatures from T_{cl} to $(T_{cl} - 20)°C$, and suggest that heat production increases linearly as temperature falls, whereas the experimental evidence suggests that the rate of increase is slower at temperatures just below T_{cl} but that this rate increases at still lower temperatures. Values for the amounts of meal which are equivalent to the increases in heat production have been calculated assuming that the meal contains 12 MJ ME kg^{-1} and that ME is utilised for thermoregulatory heat production with an efficiency of 100% (Verstegen *et al.*, 1973); these are also presented in *Table 4.11*. The values for groups of pigs are in reasonable agreement with those calculated by Verstegen *et al.* (1973).

At temperatures higher than the values for T_{cu} presented in *Table 4.9* the pig comes under increasingly severe heat stress; rectal temperature increases and respiratory rate also increases. In association with these changes it has been shown that the rate of heat production is increased and consequently the rate of energy retention is decreased (Holmes, 1973; 1974).

Table 4.11 Calculated values[1] for the rate of increase in heat production for a decrease of 1°C in air temperature below T_{cl} and for the equivalent weight of meal (*see text for calculations*)

Live weight (kg)	Increase in heat production for a decrease of 1°C in air temperature below T_{cl} (kJ per pig daily)	Amount of meal per pig equivalent to the increase in heat production (g per day)
INDIVIDUAL PIGS		
2	47	4
20	163	14
60	316	26
100	430	36
140 (Fat)	408	34
140 (Thin)	710	59
GROUPS OF PIGS		
20	160	13
60	304	25
100	417	35

[1]Calculated for controlled climatic conditions; see Table 4.5

Protein metabolism

Fuller and Boyne (1971, 1972) found that the heat production of individual pigs, 20–90 kg live weight, was increased at air temperatures of 5° and 13°C compared with 23°C; in association with these changes in energy metabolism the retention of nitrogen was also reduced at the lower temperatures and accompanied by an increase in the urinary excretion of nitrogen. A similar increase in urinary nitrogen excretion was shown to occur in sows exposed to 6–8°C (Holmes and McLean, 1974) in association with an increase in heat production. However, Verstegen et al. (1973) found no such effect of exposure to 8°C on the nitrogen metabolism of a group of pigs, 30–40 kg live weight, although heat loss was increased at 8°C when compared with 20°C. Holmes (1973, 1974) found increases in the excretion of nitrogen in the urine of pigs exposed to 33°C, compared with 25°C, with corresponding decreases in nitrogen retention.

Food intake

Heitman and Hughes (1949) found that as air temperature increased above 21°C, rectal temperature and respiratory rate increased and food intake decreased. Below 21°C, however, food intake increased with little change in the other measurements. In view of the association between respiratory rate and food intake under hot conditions it may be concluded that respiratory rate alone can serve as a useful index of heat stress; if it is above 50–60 per minute for a resting pig, the pig is under some degree of stress. The decreased intake of food at high temperatures may necessitate increased concentrations of certain components in the food; for example, vitamins and minerals (see below for thiamin requirements). It is of interest to note that Holme and Coey (1967) found that pigs fed *ad lib.* and kept at 12°C ate more food than those kept at 22°C to such an extent that they grew faster. Nevertheless, the utilisation of food for gain in live weight was less efficient at the lower temperature.

Carcase composition

Sorenson (1961) quoted results which showed that carcase back-fat depth increased at temperatures of 24°C and 3°C compared with intermediate temperatures; however the number of animals used was small. Holme and Coey (1967) found that there were no differences in back-fat depth between pigs kept at 12°C and 22°C, and fed on either a restricted level or *ad lib;* carcase length was however increased at the higher temperature. Fuller and Boyne (1971) fed increased amounts of food at 5°C and 13°C so that the pigs grew at the same rate as those kept at 23°C; the pigs kept at 5°C had fatter carcases than those kept at 23°C. If the pigs were fed the same amount at all temperatures, so that growth rate was reduced at 5°C, there was no difference in carcase composition. Holmes (1971b) showed that back-fat depth was increased in pigs exposed to 33°C, compared with 25°C, with similar levels of intake at both temperatures; carcase length was increased at the higher temperature in agreement with the findings of Holme and Coey (1967). However, in cases where food intake is reduced at the higher temperature, carcase fatness is also likely to decrease (Sugahara et al., 1970). Due to the possibility that the distribution of fat within the carcase is influenced by environmental temperature, some caution is necessary when interpreting differences in back-fat depths which occur between different environmental conditions in terms of carcase composition.

In addition to changes in carcase composition, ambient temperature has been shown to cause changes in the type of fat deposited in the pig's carcase (Fuller, Duncan and Boyne, 1974).

Growth rate

If the retention of energy and protein is affected by climate, then growth rate should also be affected independently of any effect on food intake. Such effects have been reported in many trials and some of the results have been summarised by Mount (1968) and by Fuller and Boyne (1971). The latter authors reported that over the range in temperature of 13—20°C to 3—8°C growth rate decreased by 9—19 g per day per °C fall in several experiments.

The increase in heat production, and consequent decrease in energy retention, can be calculated from the results in *Tables 4.8* and *4.11*. In addition, values have been calculated from the data of Thorbek (1975) for the energy content of the gain in live weight for pigs of several live weights. These range from 5.7 MJ kg^{-1} gain for a 20 kg pig gaining at 0.30 kg per day, through 19.4 MJ kg^{-1} for a 60 kg pig gaining 0.67 kg per day, up to 23.2 MJ kg^{-1} for an 80 kg pig gaining 0.78 kg per day.

An example will illustrate the calculations involved. From *Table 4.11* it can be seen that for an individual pig of 60 kg live weight heat loss increases by 316 kJ per day per °C decrease in temperature between the lower critical temperature and a temperature 20°C lower. If this results in proportionately less fat and lean being deposited this should represent a decrease in live weight gain of:

$$\frac{316}{19.4} = 16 \text{ g per day for a fall of } 1°C \text{ at temperatures below } T_{cl}$$

(19.4 kJ g^{-1} is the appropriate value for the energy content of tissue deposited by a 60 kg pig (Thorbek, 1975)). If, However, the effect is solely on the rate of fat deposition, the increase in heat production of 316 kJ should represent a decrease in live weight gain of:

$$\frac{316}{39.7} = 8 \text{ g per day for a fall of } 1°C \text{ at temperatures below } T_{cl}$$

70 *The influence of climatic variables on aspects of productivity in pigs*

(39.7 kJ g^{-1} is the value for the energy content of fat (Brouwer, 1965)). The discrepancy between the results of these two calculations shows the importance of identifying the source of the extra heat produced in cold climates, when trying to relate this to gain in live weight.

A series of calculations, similar to those above, have been made for pigs at three live weights assuming that the extra heat was derived either at the expense of fat deposition alone, or at the expense of all tissues in the ratios in which they were deposited. These calculations were made from the data in *Table 4.11* and those of Thorbek (1975), and the results are presented in *Table 4.12*. They show that there are considerable discrepancies between the methods of calculation and that these discrepancies are greater for younger pigs because these pigs are depositing a much lower proportion of fat in their live weight gain than older pigs. These values suggest that although the effect of age and live weight of the pig on its responses to changes in climate in terms of heat losses may not be great, younger pigs are likely to be more sensitive to lower temperatures as judged by changes in live weight gain because of their greater relative deposition of protein compared with older pigs.

Mineral, vitamin and water requirements

Limited evidence is available on these topics. Peng and Heitman (1974) found that, as judged from certain biochemical techniques, the thiamin requirement of pigs growing from 28 to 90 kg live weight increased at 30°C and 35°C when compared with 20°C when expressed as mg thiamin per kg diet; at least some of this effect was probably due to the decrease in food intake at higher temperatures.

Holmes and Grace (1975) found that the urinary excretion of Ca and K increased in pigs exposed to 33°C when compared with pigs exposed

Table 4.12 Calculated values[1] for the decrease in live weight gain for individual pigs caused by a decrease of 1°C between lower critical temperature and a temperature 20°C below that (g per day per °C fall in temperature)

	Live weight (kg)		
	20	60	100
INDIVIDUAL PIGS			
(a)	4	8	11
(b)	29	16	19
GROUPS OF PIGS			
(a)	4	8	11
(b)	28	16	18

[1]Calculated for controlled climatic conditions; *see Table 4.5*
(a) Calculated by assuming that only fat deposition affected at lower temperatures
(b) Calculated by assuming that all tissues being deposited by the pig affected equally at lower temperatures

to 25°C; the increase in K excretion was associated with an increase in N excretion at 33°C. The increase in Ca excretion may be linked in some way with the lameness which occurred in pigs on a moderate level of feeding at 33°C, although Ca retention was unaffected.

The water requirements of growing pigs have been shown to increase at temperatures of 30–33°C (Mount et al., 1971).

References

Bond, T.E., Heitman, H. and Kelly, C.F. (1965). 'Effects of increased air velocities on heat and moisture loss and growth of swine.' *Trans. Am. Soc. agric. Engrs*, **8**, 167

Bond, T.E., Kelly, C.F. and Heitman, H. (1952). 'Heat and moisture loss from swine.' *Agric. Engng*, **33**, 148

Bond, T.E., Kelly, C.F. and Heitman, H. (1959). 'Hog house air conditioning and ventilation data.' *Trans. Am. Soc. agric. Engrs*, **2**, 1

Brody, S. (1945). *Bioenergetics and Growth*. Reinhold Publishing Corp., New York

Brouwer, E. (1965). 'Report of subcommittee on constants and factors.' In: *Energy Metabolism*. Ed. by K.L. Blaxter. Academic Press, New York and London

Barlacu, G., Baia, G., Ionila, Dumitra, Moisa, Donia, Tascenco, V., Visan, I. and Stoica, I. (1973). 'Efficiency of the utilisation of the energy of food in piglets after weaning.' *J. agric. Sci.*, **81**, 295

Close, W.H. (1970). 'Nutrition–environmental interactions of growing pigs.' PhD thesis, Queen's University, Belfast

Close, W.H. and Mount, L.E. (1975). 'The rate of heat loss during fasting in the growing pig.' *Br. J. Nutr.*, **34**, 279

Close, W.H. and Mount, L.E. (1976). Unpublished

Forbes, T.J. and Walker, N. (1968). 'The utilisation of wet feed by bacon pigs with special reference to pipe-line feeding.' *J. agric. Sci.*, **71**, 145

Fuller, M.F. and Boyne, A.W. (1971). The effects of environmental temperature on the growth and metabolism of pigs given different amounts of food. I. *Br. J. Nutr.*, **25**, 259

Fuller, M.F. and Boyne, A.W. (1972). The effects of environmental temperature on the growth and metabolism of pigs given different amounts of food. II. *Br. J. Nutr.*, **28**, 373

Fuller, M.F., Duncan, W.R.H. and Boyne, A.W. (1974). 'Effect of environmental temperature on the degree of unsaturation of depot fats of pigs given different amounts of food.' *J. Sci. Fd Agric.*, **25**, 205

Heitman, H. and Hughes, E.H. (1949). 'The effects of air temperature and relative humidity on the physiological well being of swine.' *J. anim. Sci.*, **8**, 171

Holme, D.W. and Coey, W.E. (1967). 'The effects of environmental temperature and method of feeding on the performance and carcase composition of pigs.' *Anim. Prod.*, **9**, 209

Holmes, C.W. (1966). 'Studies on the effects of environment on heat losses from pigs.' PhD thesis, Queen's University, Belfast

Holmes, C.W. (1971a). Growth of pigs fed cool whey at two ambient temperatures. *Anim. Prod.*, **13**, 1

Holmes, C.W. (1971b). 'Growth and back-fat depth of pigs kept at a high ambient temperature.' *Anim. Prod.*, **13**, 521

Holmes, C.W. (1973). 'The energy and protein metabolism of pigs growing at a high ambient temperature.' *Anim. Prod.*, **16**, 117

Holmes, C.W. (1974). 'Further studies on the energy and protein metabolism of pigs growing at a high ambient temperature, including measurements with fasting pigs.' *Anim. Prod.*, **19**, 211

Holmes, C.W. and Breirem, K. (1973). 'A note on the heat production of fasting pigs in the range 16 to 96 kg live weight.' *Anim. Prod.*, **18**, 313

Holmes, C.W. and McLean, N.R. (1974). 'The effect of low ambient temperatures on the energy metabolism of sows.' *Anim. Prod.*, **19**, 1

Holmes, C.W. and McLean, N.R. (1976). 'The heat production of groups of young pigs exposed to reflective or non-reflective surfaces on walls and ceiling.' In preparation

Holmes, C.W. and Grace, N.D. (1975). 'A note on the metabolism of Ca, P, Mg, Na and K by pigs growing at a high ambient temperature.' *Anim. Prod.*, **21**, 341

Ingram, D.L. (1965a). 'The effect of humidity on temperature regulation and cutaneous water loss in the young pig.' *Res. vet. Sci.*, **6**, 9

Ingram, D.L. (1965b). 'Evaporative cooling in the pig.' *Nature*, **207**, 415

Jenkinson, G.M., Young, L.G. and Ashton, G.C. (1967). 'Energy metabolism and body composition of weanling pigs.' *Can. J. anim. Sci.*, **47**, 217

Jordan, J.W. (1974). 'The effect of calorie/protein ratio on the utilisation of energy for fat and protein synthesis by the early weaned pig.' In *Energy Metabolism of Farm Animals*. Ed. by K.H. Menke, H-J Lantzsch and J.R. Reichl. Stuttgart; Universität Hohenheim

Jordan, J.W. and Brown, W.O. (1970). 'The retention of energy and protein in the baby pig fed on cow's milk.' In *Energy Metabolism of Farm Animals*. Ed. by A. Schürch and C. Wenk. Zurich; Juris Druck and Verlag

Kielanowski, J. and Kotarbinska, M. (1970). 'Further studies on energy metabolism in the pig.' In *Energy metabolism of Farm Animals*. Ed. by A. Schürch and C. Wenk. Zurich; Juris Druck and Verlag

Mount, L.E. (1960). 'The influence of huddling and body size on the metabolic rate of the young pig.' *J. agric. Sci.*, **55**, 101

Mount, L.E. (1963). 'The thermal insulation of the newborn pig.' *J. Physiol.*, **168**, 698

Mount, L.E. (1964). 'Radiant and convective heat loss from the newborn pig.' *J. Physiol.*, **173**, 96

Mount, L.E. (1966). 'The effect of wind-speed on heat production in the newborn pig.' *Q. Jl exp. Physiol.*, **51**, 18

Mount, L.E. (1967). 'The heat loss from newborn pigs to the floor.' *Res. vet. Sci.*, **8**, 175

Mount, L.E. (1968). *The Climatic Physiology of the Pig*. London; Arnold

Mount, L.E. (1974). 'The concept of thermal neutrality.' In *Heat Loss from Animals and Man.* Ed. by J.L. Monteith and L.E. Mount. London; Butterworths

Mount, L.E. (1975). 'The assessment of thermal environment in relation to pig production.' *Live Stk. Prod. Sci.*, **2**, 381

Mount, L.E., Holmes, C.W., Close, W.H., Morrison, S.R. and Start, I.B. (1971). 'A note on the consumption of water by the growing pig at several environmental temperatures and levels of feeding.' *Anim. Prod.*, **13**, 561

Morrison, S.R., Bond, T.E. and Heitman, H. (1968). 'Effect of humidity on swine at high temperature.' *Trans. Am. Soc. agric. Engrs*, **11**, 526

Morrison, S.R., Heitman, H. and Givens, R.L. (1975). 'Effect of diurnal air temperature cycles on growth and food conversion in pigs.' *Anim. Prod.*, **20**, 287

Peng, C-L and Heitman, H. (1974). The effect of ambient temperature on the thiamin requirement of growing–finishing pigs. *Br. J. Nutr.*, **32**, 1

Sørenson, P.H. (1961). 'Influence of climatic environment on pig performance.' In: *Nutrition of Pigs and Poultry.* Ed. by J.T. Morgan and D. Lewis. London; Butterworths

Sugahara, M., Baker, D.H., Harmon, B.G. and Jensen, A.H. (1970). 'Effect of ambient temperature on performance and carcase development in young swine.' *J. anim. Sci.*, **31**, 59

Stephens, D.B. (1971). 'The metabolic rates of newborn pigs in relation to floor insulation and ambient temperature.' *Anim. Prod.*, **13**, 303

Stephens, D.B. and Start, I.B. (1970). 'The effects of ambient temperature, nature and temperature of floor and radiant heat on the metabolic rate of the newborn pig.' *Int. J. Biomet.*, **14**, 275

Thorbek, G. (1974). 'Energy metabolism in fasting pigs at different live weights as influenced by temperature.' In *Energy Metabolism of Farm Animals.* Ed. by K.H. Menke, H-J Lantzsch and J.R. Reichl. Stuttgart; Universität Hohenheim

Thorbek, G. (1975). 'Studies on energy metabolism of growing pigs.' No 424 Beretning Fra Staten Husdyrbrugs Forsog, Kobenhavn

Verstegen, M.W.A. (1971). 'Influence of environmental temperature on energy metabolism of growing pigs housed individually and in groups.' Thesis 71–2. Mededelingen Landbouwhogeschool, Wageningen, The Netherlands

Verstegen, M.W.A. and van der Hel, W. (1974). 'The effects of temperature and type of floor on metabolic rate and effective critical temperature in groups of growing pigs.' *Anim. Prod.*, **18**, 1

Verstegen, M.W.A., Van Es, A.J.H. and Nijkamp, H.J. (1971). 'Some aspects of energy metabolism of the sow during pregnancy.' *Anim. Prod.*, **13**, 677

Verstegen, M.W.A., Close, W.H., Start, I.B. and Mount, L.E. (1973). 'The effects of environmental temperature and plane of nutrition on heat loss, energy retention and deposition of protein and fat in groups of growing pigs.' *Br. J. Nutr.*, **30**, 21

5

CLIMATIC ENVIRONMENT AND PRACTICAL NUTRITION OF THE GROWING PIG

D. G. FILMER
Dalgety Crosfields Ltd., Bristol

M. K. CURRAN
Wye College, University of London

Introduction

Due to the recently escalating costs of all sources of energy (whether from food or from fossil fuels), energy conservation and improved energetic efficiency are now of considerable financial importance to the pig farmer.

Fortunately, in the last few years, the realisation that food energy and fuel energy can be partially interchangeable has been demonstrated in both the pig and poultry sectors. Practical farmers have not been slow to adopt new practices involving improved environments in order to increase their financial efficiency. A good deal of the fundamental work with pigs in this country has been carried out at the ARC Institute of Animal Physiology, Babraham, Cambridge, and has been presented in the previous chapter. This chapter will concentrate on those aspects of particular interest to the pig farmer and his feed supplier.

Comparison of Energy Systems

There are various methods currently being used at research centres and within the feed trade to describe energy relationships in pig nutrition. This has led to some confusion, particularly as 1976 sees the start of metrication for the feed industry, coupled with a move towards SI units. Since this will involve a change from calories to megajoules and pounds to kilogrammes a few comments will be made about these alternative units.

Figure 5.1 shows the classical breakdown of the gross energy of the food via digestible energy (DE) and metabolisable energy (ME), to net energy (NE) which can be used for maintenance or for productive

76 *Climatic environment and practical nutrition of the growing pig*

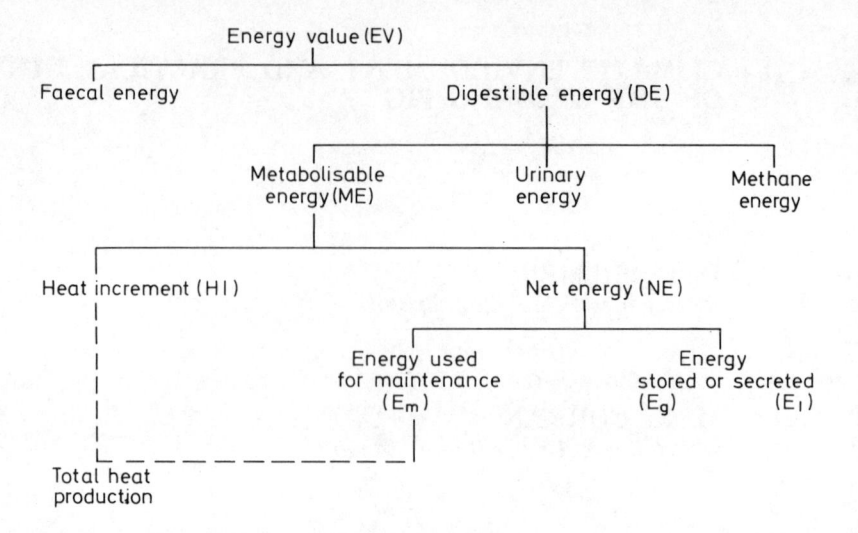

```
                        Energy value (EV)
                    ┌───────────┴───────────┐
        Faecal energy                  Digestible energy (DE)
                              ┌────────────┬────────────┐
                    Metabolisable      Urinary       Methane
                    energy (ME)        energy        energy
              ┌───────────┴───────────┐
    Heat increment (HI)           Net energy (NE)
        |                     ┌───────────┴───────────┐
        |              Energy used              Energy
        |              for maintenance          stored or secreted
        |                 (Eₘ)                  (E_g)         (E_l)
        |                     |
        L_____|
    Total heat
    production
```

Figure 5.1 Partitioning of food energy within the animal

purposes. The energy used for maintenance, together with that of the heat increment, is given off by the animal as total heat production.

It is worth stating that the practical objective of any energy system is to enable us to predict the performances of animals from a knowledge of the foods fed together with the amounts eaten.

A satisfactory energy system must take into account:

1. the large variety of ingredients available;
2. the effects of various combinations of these ingredients together with the level of feeding;
3. variations in the environment; and
4. differences in the inherent productive capacities of the animals.

A satisfactory system should therefore:

1. enable pig performance to be predicted once the above factors have been defined;
2. enable a ration to support a given level of performance to be formulated; and
3. enable the relative values of feedstuffs for a given purpose to be accurately assessed.

There are arguments that energy systems which describe the productive capacity of foods when fed are of more relevance to practical situations, the Starch Equivalent system (still used by some members of the feed trade) being typical.

Total digestible nutrients (TDN) is the measure of energy currently widely used in pig nutrition in Great Britain and the United States. It has been relatively simple to determine, and there are considerable data on raw material values. TDN takes into account the high energy value of fat, but digested protein and carbohydrates are given equal weight. Hence the protein might be considered to be deaminated and the TDN unit is somewhat more akin to a measure of ME than DE, although strictly it is analogous to neither.

The ME system which has recently had considerable support in both pig and poultry nutrition, and which has now been officially adopted for cattle, is a distinct improvement on Starch Equivalent or TDN, in that it allows the different efficiencies with which energy is used for maintenance of the various aspects of production to be accounted for. However, although it describes that part of the energy which is useful to the animal, it does not properly reflect the energy content of the food fed. This is because the amount of energy lost in the urine is not a constant and depends very much on the daily protein intake in relation to the daily protein needs of the animal for growth, and the maintenance of body protein. Thus less ME would be available in a diet where protein was excessive because of the energy losses involved in deaminating and excreting surplus protein in the urine.

Of course, in calculating ME, the amount lost in the urine must be deducted, and for practical purposes, many have assumed that urinary losses are proportional to DE intake. Assumptions of ME/DE vary from 0.92 to 0.96. However, for more accurate calculations, it is important to take the protein relationships into account and to calculate the ME of the diet from a knowledge of the DE intake and the protein deaminated. ME values of raw materials are thus not strictly additive when foods are combined into a diet because urinary losses vary with nutrient balance, level of feeding and the stage of growth and genetic potential of the pig.

If it is argued, on the other hand, that pigs are usually given well balanced diets in which urinary losses are reasonably constant, then variations in the ratio ME/DE should be small. In these circumstances it would therefore be equally satisfactory to use the more easily determined DE values. For these reasons the use of DE as the best description of food values for pigs is preferred.

A rule of thumb conversion from TDN to DE in kcal per lb, is to multiply the TDN figure by 2000 (Crampton, Lloyd and Mackay, 1957). Since the digestible energy is simply the gross energy of the feed less the heat of combustion of the corresponding faeces, it is more desirable to determine DE direct than to use the conversion. However, until a full set of raw material DE values are available, the rule of thumb conversion can be regarded as a simple stop-gap measure.

Table 5.1 shows the value of some common raw materials based on conventional TDN figures. DE in kcal per lb is simply the TDN figure multiplied by 2000 whilst ME is the DE figure multiplied by 0.916 (Diggs *et al.*, 1959).

78 *Climatic environment and practical nutrition of the growing pig*

Table 5.1 Comparison of energy units

	TDN (%)	DE^1 (kcal per lb)	ME^2 (kcal per lb)	DE^3 (MJ per kg)
Barley	71	1420	1300	13.10
Maize	78	1560	1430	14.40
Wheat	81	1620	1485	14.95
Oats	62	1240	1135	11.45
Wheat bran	65	1300	1190	12.00
White fishmeal	62	1240	1135	11.45
Meat meal	79	1580	1445	14.55
Soyabean meal	73	1460	1335	13.45
Tallow	230	4600	4215	42.45
Normal diet	68	1360	1245	12.55

[1] 100% TND ≏ 2000 kcal per lb DE (Crampton, Lloyd and Mackay, 1957)
[2] ME = 0.916 DE (Diggs *et al.*, 1959)
[3] 1 cal = 4.184 joules
∴ 1 kcal = 4184 joules
   1 lb = 0.4536 kg

So, 1 kcal per lb = $\frac{4184}{0.4536}$ J per kg = 9224 J per kg or 0.009224 MJ per kg

Metrication

Both kcal and lb are now replaced by megajoules (MJ) and kilogrammes (kg). The conversion is a simple mathematical calculation as 1 cal = 4.184 joules and 1 lb = 0.4536 kg. One kcal per lb therefore = $\frac{4.184}{0.4536} \times 10^{-3}$ = 0.009224 MJ kg^{-1}.

The fourth column on the table therefore is derived from the second column multiplied by this factor. It is a mathematical quirk of the data that DE expressed in megajoules per kilogram is similar to ME expressed in kcal per lb with the decimal being shifted two places.

Energy Calculations

All energy balance calculations should start with an assessment of the daily intake of digestible energy. It is commonly assumed that the digestible energy contents of raw materials are simply additive in mixed diets. However, there is evidence with fats, that the fatty acid configuration modifies the degree of absorption through the small intestine. Fatty acid configuration therefore must be taken into account, particularly where fat is a major source of energy in the diet.

There is some evidence that increased environmental temperatures in the piggery increase the digestibility of energy in the food. Fuller and Boyne (1972) showed that from 5°C to 23°C, digestibility of energy increased by 2.1% (or 0.12% per °C rise in temperature). This difference, though small, was statistically significant and should be taken into account when assessing the economics of different environmental temperatures. Incidentally, digestibility of energy was not affected by level of daily food intake. Nor was mean urinary energy loss or methane production affected by temperature.

Daily digestible energy intake is clearly affected by the energy concentration in the diet, together with the daily quantities of food fed (*Table 5.2*).

Although fully accepted in poultry nutritional circles, there has been some reluctance to accept that a given level of digestible energy can be provided to the pig by different combinations of dietary energy concentration and food intake. However, Curran, Filmer and Trapnell (1975) showed a direct relationship between daily nutrient intake and daily gain, *irrespective* of nutrient concentration in the diet.

Table 5.3 shows the composition of the diets involved, which differ in energy concentration by 17%. The High Nutrient Diet (HND) was fed from 18 kg to 90 kg whilst the alternative dietary regime, which conformed to ARC standards, comprised two stages. Note that the actual level of protein recovered in the ARC diet was higher than that calculated. The experiment used 192 Landrace pigs and involved two diets, six feeding scales and both sexes (hogs and gilts). All pigs were fed on a time scale and *Figure 5.2* indicates the six levels of maximum food intake per day. *Figure 5.3* shows the relationship between daily gain and daily nutrient intake. The regression of daily gain on mean daily energy intake was positive and significant for both diets. The slope of these two lines

Table 5.2 The effect of DE concentration of the diet and daily food intake on the daily DE intake

Daily DE intake (MJ)	=	*DE in diet* (MJ per kg)	X	*Daily food intake* (kg)
37.5	=	10	X	3.75
37.5	=	12.5	X	3
37.5	=	15	X	2.5

Table 5.3 Analysis of diets (air-dry basis)

	ARC 1[1]	ARC 2[1]	HND[1]
Crude protein (%)[2]	17.7	15.8	18.6
Crude protein (%)[3]	17.4	14.5	18.5
Total lysine (%)[3]	0.79	0.60	0.97
Metabolisable energy (MJ per kg)[3]	11.2	11.2	13.1

[1] For full explanation *see text*
[2] By analysis
[3] Calculated

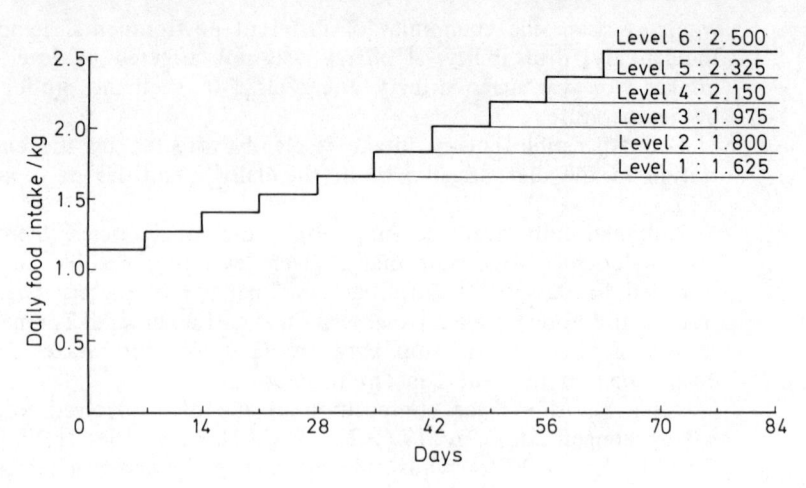

Figure 5.2 Daily food intakes. Live weight at start = 18 kg

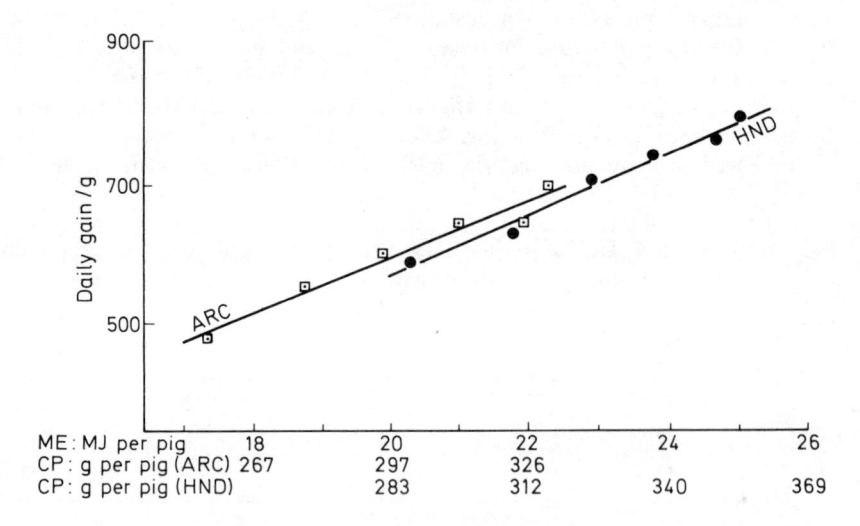

Figure 5.3 Relationship between daily gain and daily nutrient intake. CP = crude protein

and their intercepts did not differ significantly, indicating that the performance response to increased nutrient intake was independent of dietary concentration. The slight difference in intercept may be accounted for by the higher protein intake on the ARC diet compared to that which was intended.

Figure 5.4 shows that the regression of food conversion on daily nutrient intake within diets was also significant. The slope of the regressions did not differ significantly between diets indicating that regardless of nutrient density, food conversion improved as food intake increased.

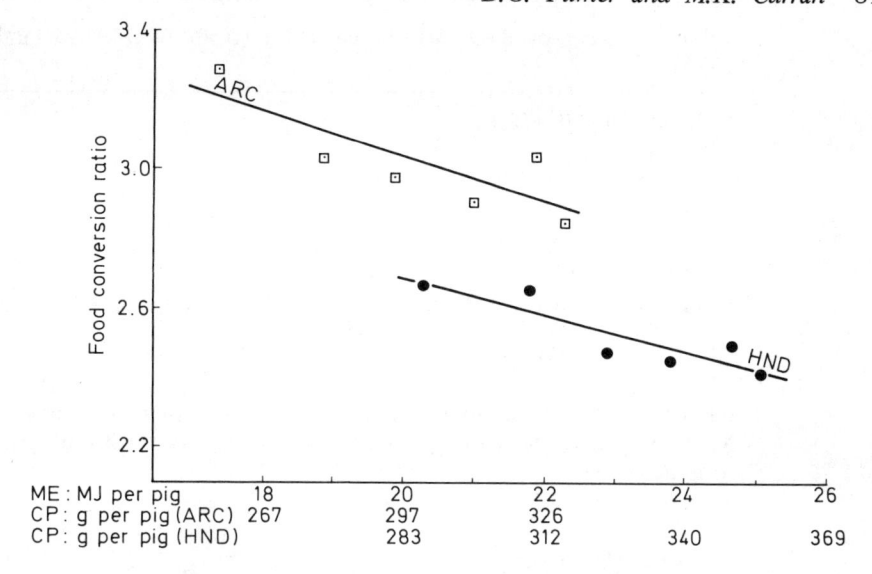

Figure 5.4 Relationship between food conversion ratio and daily nutrient intake

At corresponding intakes of energy, food conversion was better on the high density diet, as less food was required to provide these intakes.

Clearly the pig responds to a given daily intake of energy and of nutrients. Within limits, dietary composition and feeding scale can be manipulated to achieve a given required daily nutrient intake. Economically this should be done to minimise daily food costs and the optimal solution will vary from time to time depending on the relative costs of ingredients.

Partition of Energy

Once the daily DE intake has been established, the amount excreted in the urine should be calculated and subtracted. As mentioned previously, this will depend on the amount of protein deaminated. There is some loss of energy in methane from the pig, though this is not substantial, and can certainly be ignored beneath 40 kg live weight. The resultant metabolisable energy is then available for use by the pig. Recently, Whittemore and Fawcett (1976) have proposed estimates for the energy costs of these uses which are given in *Table 5.4*.

Maintenance

This clearly varies with the bodyweight of the pig, and is estimated in a thermoneutral environment. Various estimates have been derived, mainly

Table 5.4 Fate of daily ME intake (MJ) (from Whittemore and Fawcett, 1976)

1. Maintenance (0.475 $kg^{0.75}$)
2. Protein turnover
   (0.0073 × protein turnover, g)
3. Protein deposition
   (0.0236 × protein deposited, g)
4. Energy cost of cold thermogenesis
5. Lipid deposition, i.e. ME left after items 1, 2, 3 or 4 have been deducted
   (53.5 MJ per kg lipid deposited)

related to metabolic bodyweight ($kg^{0.75}$), and there is scope for refining these estimates, particularly as far as the various strains of pigs are concerned.

Protein Turnover

Protein in the body tissues is not static, but is regularly turned over in a dynamic fashion. Whittemore and Fawcett (1975, 1976) suggest that the total protein turnover is related to the rate of protein synthesis as well as the total protein mass. The energy cost of protein turnover would therefore increase as the lean body mass increases with time, and would be higher for boars (which have a higher rate of protein growth) than for gilts.

Protein Deposition

The rate of protein deposition depends on the genetic potential of the animal in relation to the protein or amino acid supply. Current thinking (*Figure 5.5*) suggests that over the main period of the growing commercial pig's life, the maximum rate of protein deposition is constant for a given animal. Should daily protein or amino acid intake be insufficient to meet the demands of this genetic maximum, then a slower rate of lean tissue growth will take place. Environmental effects are minimal except where they limit the daily supply of protein and the associated energy required. For example, high temperature could reduce protein deposition due to depressed food intake. *Figure 5.6* shows schematically the relation between daily protein intake and the rate of protein deposition. Assuming an adequate supply of energy, protein in excess of that required for maintenance is used for protein deposition with constant efficiency until the maximum genetic rate of protein deposition (Pr_{max}) is reached. Above this, excess protein intake has to be deaminated. The energy content of protein deposited is 23.6 MJ per kg protein. Once the rate of protein deposition has been estimated, the energy content of the protein deposited can thus be calculated.

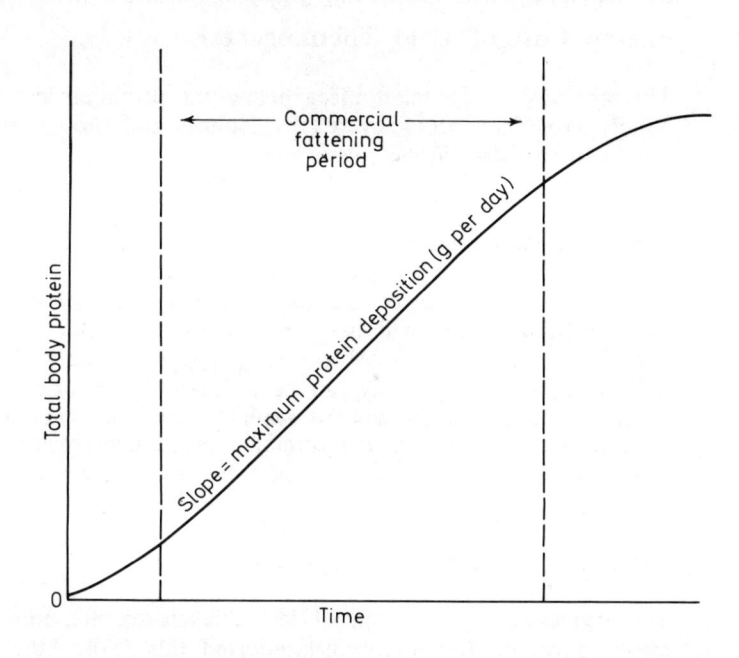

Figure 5.5 Protein growth rate (genetic potential). Typical protein growth rates: boars, 140 g per day; gilts, 110 g per day; hogs, 80 g per day

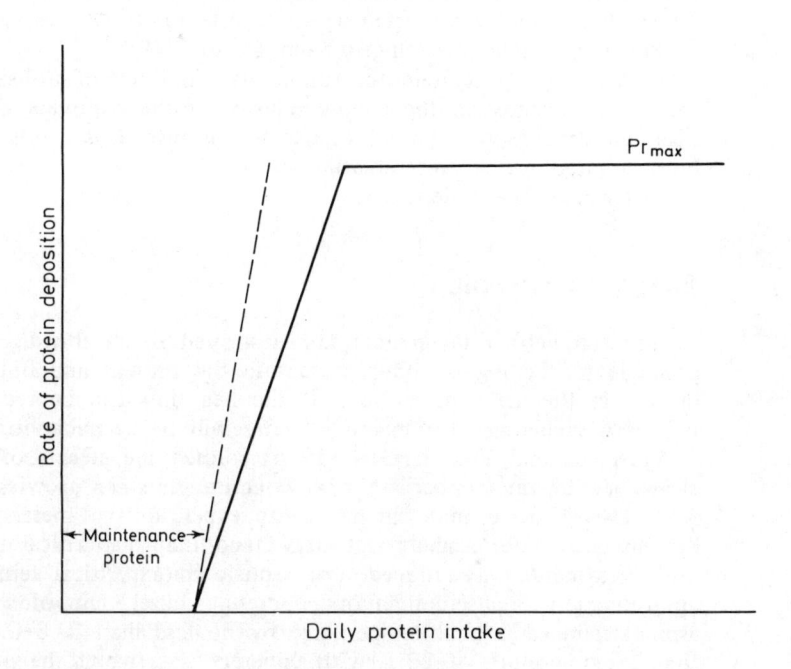

Figure 5.6 Individual pig response to protein intake

Energy Cost of Cold Thermogenesis

This obviously only takes place below the lower critical temperature. The details have been fully covered by Holmes and Close (*see* Chapter 4) but reference to this will be made later.

Lipid Deposition

Once the energy costs of the four previous items have been allowed for, the remaining energy will be deposited as lipid. The estimated energy cost (*Table 5.4*) is 53.5 MJ per kg deposited. Clearly the more the energy cost of cold thermogenesis can be reduced by attention to the environment, the more energy will be available for lipid and protein deposition. Energy saved by providing a suitable environment in these circumstances will allow for a higher level of performance on the same diet.

Efficiency of Energy Conversion

The efficiency of utilisation of ME depends on the purpose to which the ME is directed. Breirem (1939) reported this to be 80.7% for maintenance, whilst that for growth and fattening was 66.2%. In Lund's experiments (1938) utilisation for growth and fattening was 68.3% and that for gain of almost pure fat in large pigs was 71.5%. Previously Breirem (1935) had found that efficiency of utilisation of ME for growth and fattening in young pigs ranged from 65 to 70%.

In the new ME system for ruminants efficiency of utilisation of ME for cattle depends on the energy density of the complete diet. It is an open question as to whether this is so for pigs. If so, this factor should be accounted for when evaluating the economics of the use of diets of differing nutrient densities.

Energy Conservation

Energy can neither be created nor destroyed so all the digestible energy available to the pig is either present in the protein and lipid deposited, or lost in the urine or as heat. If the heat loss can be reduced by improved insulation, for example, more will be available for growth.

Verstegen and Van der Hel (1974) studied the effects of asphalt and straw bedding in comparison with concrete slats for growing pigs (*Table 5.5*). They showed that the heat output per unit of metabolic weight was decreased by both asphalt and straw, and that the critical temperature for both treatments was reduced. For asphalt floors critical temperature was approximately 3°C less than on concrete, whilst 3 cm of straw on asphalt reduced critical temperature by no less than 7–8°C. This means that a temperature of 20°C with concrete slats would be needed to give the same performance on a given daily diet compared with 12–13°C on

Table 5.5 Effects of type of floor on energy balance and growth (after Verstegen and Van der Hel, 1974)

	Asphalt		Straw		Concrete	
ME intake (MJ per kg$^{0.75}$)	1.15	1.16	1.14	1.14	1.21	1.22
Heat output (MJ per kg$^{0.75}$)	0.66	0.65	0.67	0.67	0.72	0.73
Extra thermoregulatory heat (MJ per kg$^{0.75}$)	14	31	18	15	70	47
Lower critical temp (°C)	13.7	16.1	12.3	11.3	19.9	18.6
Live weight gain (kg/day)	0.63	0.58	0.59	0.61	0.55	0.54

Table 5.6 Energy cost of cold thermogenesis (from Whittemore and Fawcett, 1976)

HEAT LOSS IN THERMONEUTRAL ZONE (MJ)

= Intake of ME $- \begin{pmatrix} 0.0236 \text{ (Protein deposited g)} \\ + 0.0393 \text{ (Lipid deposited g)} \end{pmatrix}$

LOWER CRITICAL TEMP. °C (T_{CL})

= 26.6 − (0.59 × heat loss)

Depends on insulation and probably breed

HEAT DEFICIT (HD)

= T_{CL} − house temperature

ENERGY COST OF COLD THERMOGENESIS (MJ)

= 0.016 HD × kg$^{0.75}$

straw bedding over asphalt. Live weight gain on both straw and asphalt floors was significantly greater than on concrete. The lower critical temperature depends on heat loss in the thermoneutral zone, and this can be estimated by deducting the energy content of the protein and lipid deposited daily from the ME intake. Using values suggested by Whittemore and Fawcett (1975) the lower critical temperature can then be calculated (*Table 5.6*). The more food that is fed the greater the heat increment of feeding and therefore the lower will be the lower critical temperature. Holmes and Close have provided a very useful table indicating expected lower critical temperatures for pigs at different weights at different levels of energy intake (*see Table 4.8*, p. 64).

86 *Climatic environment and practical nutrition of the growing pig*

If the pig is in an environment where the effective house temperature is lower than the lower critical temperature, there is a heat deficit (HD). This is just the difference in °C between the two.

The energy cost of cold thermogenesis can then be calculated. This extra energy cost can be met by feeding more food. Verstegen and Van der Hel (1974), using their straw/asphalt/concrete data, estimated that with a food containing 12.5 MJ ME per kg, 0.3 g extra food was required per pig daily for every 1°C below the lower critical temperature. *Table 5.7* sets this out together with the data just presented by Holmes and Close (p. 67). There is fairly close agreement between the two sets of data, but as the data from Holmes and Close are more recent, these have been used to construct *Table 5.8*. This indicates the extra food required per pig per day at 1, 5, 10 and 15°C respectively below T_{cl}, for pigs from 20 to 90 kg, live weight. This shows that approximately 0.25 lb more food is required to maintain the same growth rate when the temperature is 5°C beneath the lower critical temperature. This rises to approximately 0.50 lb when 10°C beneath T_{cl}, and climbs to nearly 1 lb per pig per day when 15°C beneath T_{cl}. This information is of value to practical pig farmers in that if they can ensure that their pigs do not fall beneath their lower critical temperature there is an effective saving of feed energy.

Dalgety Crosfields have developed an Ultraplan thermometer (*Figure 5.7*) based on the data contained in *Table 5.8*, and it indicates visually the extra food required per pig per day, when minimum temperatures become too low.

Holmes and Close (p. 70) also indicated the reduction in growth which occurs when pigs are kept beneath their lower critical temperature on the

Table 5.7 Extra food (12.5 MJ ME/kg) required per day per °C below T_{cl}

Live weight (kg)	Verstegen and Van der Hel (1974)		Holmes and Close (Table 4.11, p. 67)	
	Per pig (g)	Per kg (g)	Per pig (g)	Per kg (g)
20	6	0.3	13	0.65
60	18	0.3	25	0.41
100	30	0.3	35	0.35

Table 5.8 Extra food (12.5 MJ ME per kg) required per day below T_{cl}

Live weight (kg)	1°C	5°C		10°C		15°C	
	(g)	(g)	(lb)	(g)	(lb)	(g)	(lb)
20	13	65	0.14	130	0.29	195	0.42
40	18	90	0.20	180	0.40	270	0.60
60	25	125	0.28	250	0.55	375	0.82
90	33	165	0.36	330	0.73	495	1.09

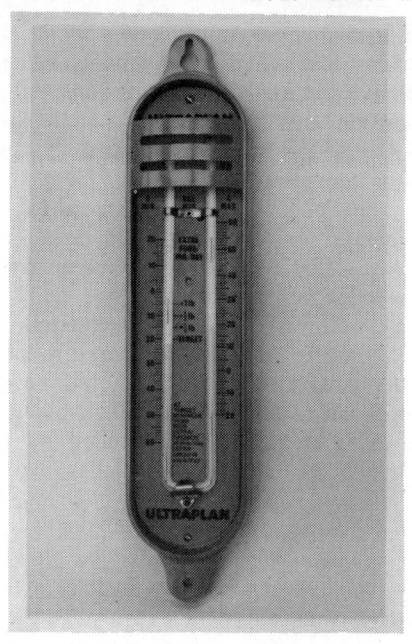

Figure 5.7 Ultraplan thermometer developed by Dalgety Crosfields

Table 5.9 Reduction in live weight gain below T_{cl}

Live weight (kg)	5°C g per d	5°C lb per week	10°C g per d	10°C lb per week	15°C g per d	15°C lb per week
Reduced growth						
20	145	2.2	290	4.5	435	6.7
60	80	1.2	160	2.5	240	3.7
100	95	1.5	190	2.9	285	4.4
Reduced fat only						
20	20	0.3	40	0.6	60	0.9
60	40	0.6	80	1.2	120	1.8
100	55	0.8	110	1.7	165	2.5

same quantity of food, and although this may not have seemed substantial when expressed in g of live weight per day per °C, *Table 5.9* shows that, for example, a 10°C drop beneath the lower critical temperature for a 20 kg pig reduces growth by 4.5 lb per week, which is quite considerable.

To summarise, the fate of available ME is either to produce:

1. an increase in energy content of the carcase (protein and lipid deposited) or,

88 *Climatic environment and practical nutrition of the growing pig*

2. heat loss.
(*a*) That lost when in thermoneutral zone, i.e. maintenance ME (activity, essential body functions, etc.) + energy cost of protein deposition and turnover.
(*b*) Any extra heat loss due to the pig being beneath the lower critical temperature.

With a knowledge of these relationships, most of which can be quantified with greater or lesser precision, it is possible to calculate, albeit inexactly, the effects of a given diet and feeding scale on pig performance. Some knowledge of the potential for daily protein deposition of the pig (which may well vary from strain to strain) is required together with information on the effective temperature of the environment.

From the equations given, it should be possible to estimate the lower critical temperature and if the effective environmental temperature is less than this, extra energy will be needed by the pig to maintain its body heat.

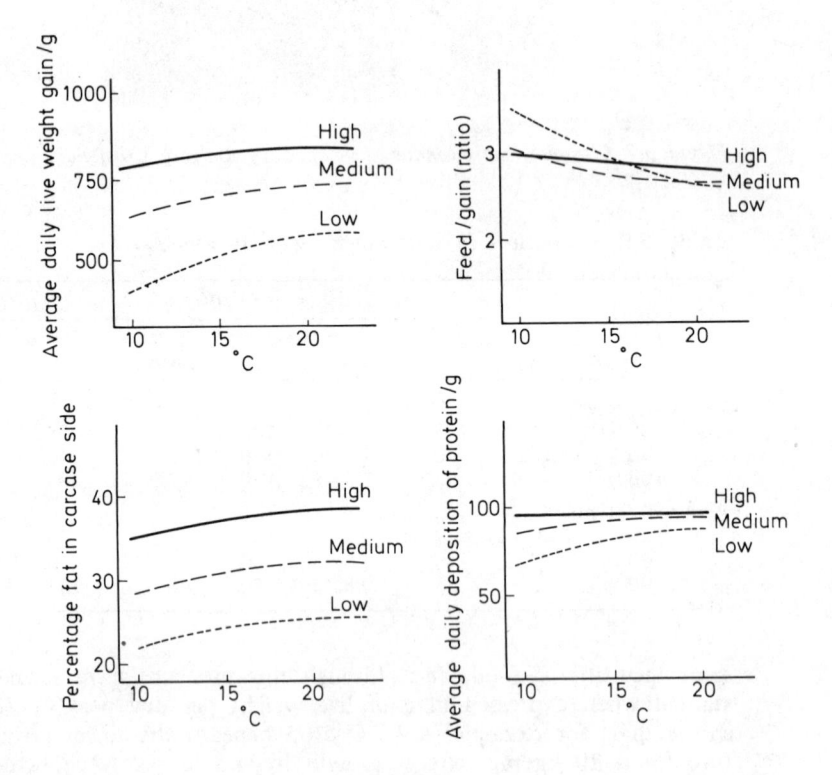

Figure 5.8 Responses to house temperature by growing pigs given three ration scales (Whittemore and Elsley, 1976, reproduced by permission of Farming Press Ltd.

	Low	Medium	High
Ration (kg) at 20 kg	0.80	0.95	1.1
Ration (kg) at 100 kg	2.5	3.2	3.8
Average daily	1.6	2.0	2.4

Some essential lipid will automatically be laid down as part of lean tissue growth, and sufficient energy is required for this. Any surplus protein in the system over that required for lean tissue growth will be deaminated and this will entail an energy loss. If energy is surplus after these various commitments have been met, it will be laid down as extra fat.

Whittemore and Fawcett (1975, 1976) have developed a computer model which utilises these above relationships to simulate mathematically the reaction of a growing pig of defined genotype to a given daily nutrient intake in a stated environment. The model calculates the anticipated daily growth of lean tissue and fat from the input data on a daily basis. It then adds the expected growth to the assumed lean tissue and fat content of the pig at the start weight, and proceeds to calculate the next day's performance. The relationships, as shown above, change with bodyweight, body composition, nutrient intake, etc., so that this iterative process may more nearly simulate the real animal than conventional calculations based on a longer time scale. It also incorporates the calculation of critical temperature and extra heat loss *if environmental temperature is suboptimal.*

This new approach is of considerable interest, but it is still too early to make detailed comment. However, like all computer systems, the output information is clearly limited by the assumptions inherent in the programme and the relevance and accuracy of the data incorporated. Nevertheless, the technique is capable of dealing with new information as it arises and its accuracy and relevance will no doubt be continually improved.

Figure 5.8 (from Whittemore and Elsley, 1976) illustrates results of simulating various house temperatures and three scales of feeding.

The simulation was from 20 to 100 kg live weight using 'pigs' with genetic potential of 100 g per d protein deposition. The diet simulated contained 13 MJ DE per kg and 150 g per kg DCP with a Biological Value of 70.

Energy/Protein Ratios

Environment, particularly house temperature and floor insulation, puts a higher or lesser demand on the pig for energy from the food. On the other hand temperature seems to have little effect on the daily protein requirement. Under favourable conditions therefore, daily energy intake from the food can be reduced, without detracting from lean tissue gain, though surplus fat in the carcase will be reduced.

A reduction in energy intake could be achieved by reducing energy in the diet, or decreasing the feed intake. The latter is likely to be more economic under present conditions, but if this course is taken it will probably become necessary to increase the protein percentage in order to maintain a satisfactory daily protein intake. Whichever course is taken, the energy/protein ratio of the diet has clearly been altered.

If environment is taken into account when formulating a pig food (as it should be), the daily intakes of both protein and energy must be treated as being of equal importance *in their own right*. It would be better if energy/protein ratios were forgotten altogether.

Carcase Yield

Clapp *et al.* (1975) showed that pigs kept at 31°C and 21°C in environmental chambers produced a higher killing out percentage than those fed in outside pens or an enclosed finishing barn. Data on this aspect are limited but if genuine increases in carcase yield are associated with better environments this must be accounted for in economic evaluations of these environments.

Carcase Quality

It is obvious from earlier discussion that environment has a considerable effect on the quantity of fat produced by the growing pig. Manipulation of energy input by level of feeding is the normal way in which practical pig farmers control fat levels in the carcase to meet grading standards.

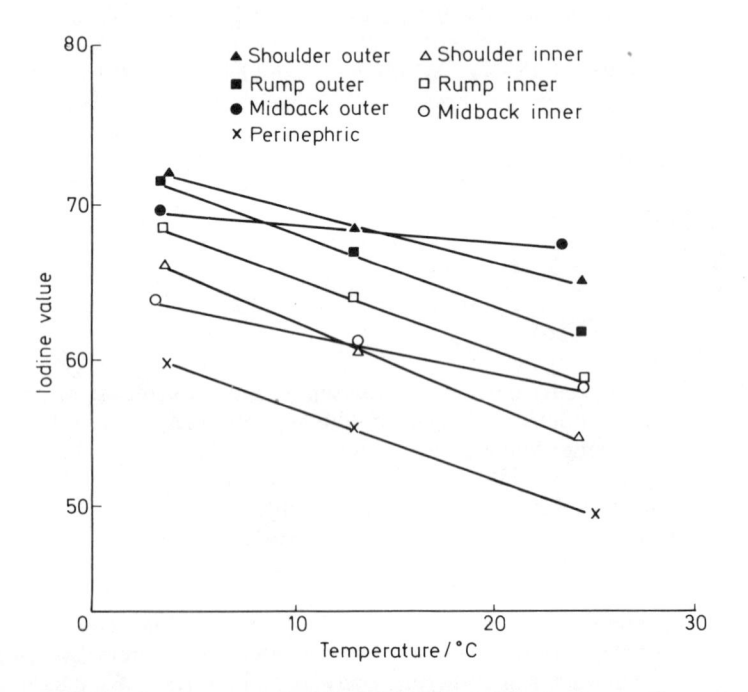

Figure 5.9 Mean iodine values of fat at different temperatures

Farmers now also appreciate that an improved environment will enable more energy to be diverted to growth and fat production. It is common experience that in good environments, food intake has to be restricted more severely than normal if over-fatness is to be avoided.

Increased protein intake has been shown to increase lean/fat ratio in pig carcases. Part of this may be because Pr_{max} (the maximum genetic rate of protein deposition) was limited by inadequate protein intake previously. However if this were not the case, the surplus protein in the system would have to be deaminated, and this requires an energy input. This will divert some of the energy that would have gone to produce surplus fat. The effect of surplus protein therefore is to reduce fat synthesis rather than increase protein deposition.

Economically of course, it would be better to reduce energy intake by restriction of food intake.

The effects of environmental temperature therefore cannot be predicted, unless full details of the nutritional inputs and the genetic capacities of the pigs are known. This could account for some of the apparently conflicting data reported in the literature. However if these details are known or can be estimated, the effects of temperature can be assessed.

Fat quality may be affected by temperature as shown by Fuller, Duncan and Boyne (1974). Harder, more saturated fats were associated with increased temperature. *Figure 5.9* shows the effects of environmental temperatures of 5, 13 and 23°C on iodine values. It is fortunate that more saturated and therefore harder fats are produced with increased environmental temperatures.

Some body organs have been shown to be affected by environment. Comberg *et al.* (1972) showed that heart weight was decreased with increasing temperatures. Possibly this was due to decreased physical activity.

Conclusions

Sufficient quantitative data are now available for some aspects of the climatic environment to be taken into account when formulating growing pig diets. No doubt the accuracy and extent of these data will be increased over the next few years. Energy considerations are affected primarily, higher environmental temperatures releasing more food energy for productive purposes. Provided requirements for other nutrients such as protein, amino acids, vitamins, etc., are considered on a daily basis, there is little information yet to indicate that environment has much influence on these.

It is suggested that DE should be the preferred unit for describing the energy value to the pig of feedingstuffs and raw materials. However, if quantitative estimates of lean and fat deposition arising from feeding a given diet are required, ME must firstly be derived, taking energy lost in the urine into account. This will depend on the daily amino acid supply in relation to the lean body mass and the genetic potential of the pig for lean tissue growth, together with the adequacy of the energy supply relative to the environment.

There is scope for mechanising the somewhat complex calculations involved in order to produce optimally economic feeding solutions, involving nutritional and environmental factors as well as genetic and market considerations.

References

Breirem, K. (1935). *Beretn. Forsøgslab* no. 162
Breirem, K. (1939). *Tierernährung*, **11**, 487
Clapp, K.L., Ramsey, C.B., Tribble, L.F. and Gaskins, C.T. Jr. (1975). Texas Tech. University, Lubbock
Comberg, G., Plischke, R., Wegner, W. and Feder, H. (1972). *Znechtungskunde*, **44** (2), 91
Crampton, E.W., Lloyd, L.E. and Mackay, V.G. (1957). *J. Anim. Sci.*, **16**, 541
Curran, M.K., Filmer, D.G. and Trapnell, M.G. (1975). Proc. *Br. Soc. Anim. Prod.*, **4**, 117
Diggs, B.G., Becker, D.E., Terrill, S.W. and Jensen, A.H. (1959). *J. Anim. Sci.*, **18**, 1492
Fuller, M.F. and Boyne, A.W. (1972). *Br. J. Nutr.*, **28**, 373
Fuller, M.F., Duncan, W.R.H. and Boyne, A.W. (1974). *J. Sci. Fd Agric.*, **25** (2), 205
Lund, A. (1938). Heretn. Forsøgslab no. 180
Verstegen, M.W.A. and Van der Hel, W. (1974). *Anim. Prod.*, **18**, 1
Whittemore, C.T. and Elsley, F.W.H. (1976). *Practical Pig Nutrition*. Farming Press Ltd., Ipswich
Whittemore, C.T. and Fawcett, R.H. (1975). Proceedings of Seminar on Computer Optimisation of Pig Feeding Programmes, University of Edinburgh
Whittemore, C.T. and Fawcett, R.H. (1976). *Anim. Prod.*, **22**, 87

6

THE NUTRITION OF RABBITS

J. PORTSMOUTH*
RHM Agriculture Ltd., Berkshire

The data presented here will deal only with the rabbit raised for meat production and not the rabbit bred and grown for either show (fancy) or Angora wool.

The Rabbit Industry

In 1962 it was estimated that Great Britain was producing some 40 000 tons of rabbit meat. At that time only 1 rabbit in 10 was produced under so called modern management methods and the remaining 9 were grown for the table by backyard producers who used extensive management systems and bulk feeding stuffs. Prior to the first major epidemic of myxomatosis in 1953 some 95 000 tons of rabbit meat were produced in Great Britain (Portsmouth, 1962). By 1972 it was estimated that production had fallen to 15 000 tons with a further 10 250 being imported, mainly from China (Sinquin, 1973).

According to Parkin (1975) the output in Great Britain fell to some 6500 tons in 1974, with importation of 8490 tons 90% of which came from China and 10% from Australia. *Table 6.1* summarises the position.

Table 6.1 Rabbit meat production and imports 1950 to 1974 (metric tonnes)

	1950	*1962*	*1972*	*1974*
Home produced	49 200	39 350	14 760	6400
Imports	44 250	*	10 086	8354

*None available

The Department of Trade and Industry stated that the annual per capita consumption of rabbit meat in 1975 was 318 g (0.7 lb). This compares with 1.6 kg before the Second World War and 12.27 kg for poultry meat. *Table 6.2* shows the comparative annual consumption of rabbit, chicken, turkey and beef.

*Present address: Peter Hand (GB), Russell House, Rickmansworth, Herts.

94   *The nutrition of rabbits*

Table 6.2 1973 Consumption of rabbit and competitive meats per capita (kg)

Rabbit	0.32
Chicken	12.50
Turkey	1.50
Beef (on bone)	16.60

In 1961 the industry produced approximately 2 million broiler rabbits and some 15 million extensive (bulk fed) meat rabbits. According to MAFF (Parkin, 1975) it was estimated that the number of does in the UK in 1974 was around 180 000. Based on an output of 36 young per doe per year the UK industry produced some 6.5 million broiler meat rabbits in 1974. Again, it is estimated that the average unit size is 80 does producing 2880 meat rabbits per annum. Based on an average doe unit of 80, there are only 2250 rabbit producers in the UK. There are, of course, a few units with over a 100 does and the odd one with several hundred. It is therefore a very small industry made up principally of small scale production units.

Based on 1974 population estimates the total amount of rabbit food required by the industry is approximately *36 000 metric tonnes* per annum.

Within the EEC France has the largest rabbit population with an estimated 9 million does in 1970 producing some 180 million meat rabbits. Italy, on the other hand, appears to have an annual output of 60 million rabbits and West Germany some 15 million. It seems that without exception the rabbit industry within the nine countries of the Common Market is in the hands of the small, less intensive, livestock farmer and small holder.

The Digestive Tract and Physiology of Digestion

The rabbit is a monogastric and herbivorous animal with a digestive system similar in some respects to that of the pig, cow and horse. A comparison between these three species is shown in *Table 6.3*.

Table 6.3 Relative capacity of the organs of the alimentary tract in different species (% of total)

	Horse[1]	*Cow*[1]	*Pig*[1]	*Rabbit*
Stomach	9	71	29	34
Small intestine	30	19	33	11
Caecum	16	3	6	49
Colon	45	8	32	6

[1] Jacquot, Le Bars and Simmonnet (1958)

The stomach of the rabbit is allied to that of the horse which lacks power of contraction at the anterior and mid regions. Even in perfect health the rabbit stomach is more than half full. In the wild state the rabbit is a continuous feeder of small meals averaging some 5 g to 10 g at a time. It is also a 'chewer' of food and only if restricted in amount by volume and time does it bolt its food. In the wild state, bolted and undigested food has been alleged to cause mucoid enteritis in growing rabbits. Enteritis is the most serious and common cause of mortality in the modern rabbitary.

The rabbit is an 'inverted ruminant' for it has a very large caecum which is involved in the microbial fermentation of fibrous material, and yet the intestine which absorbs the nutrients is sited distal to the caecum whilst the cow's rumen is proximal to the intestine. To overcome this apparent disadvantage the rabbit practices *coprophagy* or pseudorumination. It has also been called caecotrophy.

COPROPHAGY

The term 'coprophagy' is used to describe a process whereby the rabbit removes a soft type of faeces direct from its anus. This is recycled in the digestive system. The practice of coprophagy is a natural one and is also practised by the rat. It does not indicate a nutritional deficiency but the reasons for it are of immense interest. Blount (1945) observed that coprophagy did not occur in young rabbits on a milk diet and that the formation of coprophagic pellets was not prevented by giving large quantities of a vitamin B complex source. Nor was it prevented by including dairy cow faeces in the diet. He did note, however, that when food was available throughout 24 h the coprophagic pellet output was considerably reduced and less than 10% of the normal quantity was eaten. Blount suggested that coprophagy is merely a means of providing bulk for the stomach muscles to act upon because the presence of the pellets compresses the stomach remains and thus allows more nourishment to pass into the small intestine. Another theory of this eminent author is that coprophagy developed as a compensatory habit in the wild rabbit forced to remain below ground to avoid its enemies. Although supplementary vitamins and protein do not prevent coprophagy from taking place the difference in chemical analysis between the coprophagic and normal faecal pellets would certainly indicate that the rabbit derives some nutritional benefit. *Table 6.4* shows the comparative chemical analysis carried out by the author.

96 The nutrition of rabbits

Table 6.4 Composition of faeces (DM basis)

	Normal pellets (%)	Coprophagic pellets (%)
Crude protein	11.0	35.0
Ether extract	4.0	3.5
Crude fibre	35.5	14.0
Nitrogen-free extract	41.5	36.5
Ash	8.0	11.0

It can be seen that the protein value is markedly increased and the fibre is decreased. Yoshida *et al.* (1968) also showed similar differences in the protein and fibre levels.

The amino acid sparing effects of coprophagy are seen in *Table 6.5* (Ferrando *et al.*, 1970). Compared to ordinary hard faeces there is not one amino acid which is lower in the coprophagic pellets.

Table 6.5 Amino acid composition of coprophagic and normal hard faeces (g per 100 g DM)

Amino acid	Normal faeces	Coprophagic pellets
Aspartic acid	0.97	3.06
Threonine	0.54	1.79
Serine	0.45	1.34
Glutamic acid	1.006	3.30
Proline	0.54	1.28
Glycocoll	0.62	1.59
Alanine	0.58	1.80
Valine	0.63	1.69
Methionine	0.13	0.47
Isoleucine	0.53	1.28
Tyrosine	0.24	0.93
Phenylalanine	0.54	1.10
Lysine	0.60	1.57
Histidine	0.25	0.44
Arginine	0.35	0.91
Leucine	0.89	1.88

It is claimed that coprophagy also increases B-vitamins available to the rabbit (Battaglini, 1968).

It is apparent that differences in coprophagy do exist which may be due to age, plane of nutrition, time of day and feeding programme. The age by which it commences is generally stated as 5 to 6 weeks of age. Myers (1955) stated that it is controlled by the suprarenal glands and that it only appears at 3 to 4 weeks of age. Coprophagy is an essential function, and rabbits when prevented from doing it by the fitting of collars grow less well. Type of diet appears to influence the age of onset and there is a need to examine the subject with modern feeds and feeding systems and with the fast growing strains now used.

THE REQUIREMENT FOR FIBRE

The rabbit utilises fibre less efficiently than ruminants but more efficiently than single-stomached animals. The importance of fibre to the rabbit's health status and its necessity for good growth is highly debatable. In the wild state the rabbit consumes considerable quantities of long fibre but its conversion into meat is poor. By comparison the modern rabbit managed under highly intensive husbandry conditions receives much less fibre and in many cases the fibre is entirely of the short type, especially when hay is completely omitted from the diet.

It has long been alleged that a lack of fibre in the diet of adult and growing rabbits results in fur pulling and possibly cannabalism. Templeton (1955) indicated that fibre-deprived rabbits attacked eyelashes, whiskers (feelers) and body fur and that this could be prevented by supplementary hay and green food. Based on this and other evidence Templeton recommended a minimum of 15% fibre for all rabbits. Casady and Gildow (1959) also recommend a minimum level of 15%. However, the National Research Council (NRC) (1954 and 1966) make no mention whatsoever of a specific fibre need, nor do they make any tentative recommendation. Aitken and King Wilson (1962) point out that there is little evidence to help the nutritionist select an adequate crude fibre level other than to choose a level at which the rabbit thrives. They calculated that the commonly used rations contained between 10 and 16% fibre for suckling does and 15–26% for other rabbits, whilst a pelleted ration of 13% fibre, recommended for all classes of stock, was apparently satisfactory.

Mucoid enteritis is the most serious and common cause of mortality in young rabbits intensively housed. It causes acute scouring, and occurs in nearly all rabbitaries. Lack of fibrous material and/or the incorrect source of fibre have been cited as causes of the disease. However, mucoid enteritis has been reported to occur on all types of diets containing different fibre levels and, presumably, different sources although it does appear to occur less on high fibre diets and those supplemented by hay. Rabbits fed diets based on either grass meal fibre or bran fibre to give a calculated level of inclusion of 12% show a positive preference for the longer fibres of grass meal.

The younger the rabbit, the less is its ability to utilise fibre. This statement is made on the basis of two facts: first, coprophagy, the process whereby fibre is better utilised, does not normally occur before 5 weeks of age; and second, diets high in fibre give slower growth rates than diets low in fibre during the immediate weaning period up to an age at which coprophagy begins. Sandford (1957) and Besedina (1969) showed that as the fibre content of the DM increased so the digestibility of the total organic matter decreased. The digestibility coefficient was 76 at 10% fibre and declined to 61 at 20% fibre. Besedina (1969) gave digestibility coefficients for organic matter of 76.16 and 85.7 for 16.83% and 11.77% fibre respectively. Comparative values for protein were 73.6 and 82.22, NFE 80.4 and 95.3 and fat 68.5 and 48.4. Besedina (1968) compared results of 30.1, 20.5, 15.4 and 8.9% fibre levels in rabbits from mating

98 The nutrition of rabbits

to rearing of their young at three months. The fibre level had no effect on litter size and the greatest weight gain occurred with the 20.5 and 15.4% fibre levels.

THE REQUIREMENT FOR FAT AND FATTY ACIDS

The majority of commercially pelleted feeds contain between 2 and 4% total fat. There is no specific recommendation because no requirement has been established (NRC, 1966). According to Casady, Sawin and Van Dam (1971) pregnant and lactating does should receive 3.0 to 5.5% fat and non-reproducing animals 2.0 to 3.5% fat. Using a purified diet Thacker (1956) observed that 10 to 25% fat in the diet increased weight gain when compared to 5% fat. More recently, Arrington, Platt and Franke (1974), using rabbits 6–7 weeks of age, compared semi-purified diets containing 2.4 and 14% fat. A third diet, a commercial feed with 3.6% fat, was modified to provide 11.4% fat. Increasing the fat level from 2.4 to 14% caused a reduction in feed intake with an improvement in FCE, both the protein and energy needed per unit gain decreasing with increasing fat level. The increase in fat in the commercial diet also decreased feed intake and increased weight gain. Digestibility of the fat in the commercial feed (3.6% fat) was 83.6%, but increasing the fat level to 11.4% by adding 8% corn oil resulted in a digestibility coefficient of 90%. It was assumed that the added corn oil was more readily available to enzymatic action than the fat in the natural ingredients.

There is no doubt that the rabbit, both young and adult, can utilise higher levels of fat to physiological advantage. In commercial trials unweaned rabbits have been satisfactorily fed rations containing 15% lard and in other work weaned stock fed on rations containing 5 and 7.5% added tallow outgrew rabbits fed conventional diets.

The fatty acid requirements of rabbits are unknown. That rabbits may need greater quantities of unsaturated compared to saturated fatty acids has been indicated but not quantified and investigation in this direction could show some interesting responses.

THE REQUIREMENT FOR ENERGY

Rabbits will consume sufficient food to satisfy their energy requirements. Thus, the amount of food eaten is closely related to the energy level of the ration. By comparison with other farm livestock rabbits' energy requirements are relatively high. Per unit of bodyweight they need three times as much energy as cattle.

Expression of energy measurement

The energy requirement of rabbits has been expressed as the metabolisable energy in kcal per g gain, EDF (energy of digested feed in kcal per g gain) and as TDN (total digestible nutrients) in oz per pound gain. The relationship of these measurements is shown in *Table 6.6* (after Aitken and King Wilson, 1962).

Table 6.6 ME requirements for growth and approximate equivalent EDF and TDN values

Age (weeks)	8	10	12	14	16	18	20	22	24
ME (kcal per g gain)	3.9	4.7	5.5	6.3	7.1	7.9	8.7	9.5	10.3
EDF (kcal per g gain)	4.2	5.1	5.9	6.8	7.6	8.5	9.4	10.2	11.1
TDN oz per lb gain	15.2	18.3	21.4	24.6	27.7	30.8	33.9	37.0	40.2
kJ per g gain	16.687	19.665	23.012	27.006	29.706	33.054	37.216	39.748	43.095

Little work has been done with rabbits under 8 weeks of age. There is a need to recalculate the values for ME, EDF and TDN using modern feedstuffs and modern strains of rabbits and data on the energy values of some feed ingredients can be found in Aitken and King Wilson (1962) and Sandford (1957).

Energy requirements for maintenance, reproduction, lactation and growth

Maintenance On a maintenance diet the rabbit neither gains nor loses weight but it must have sufficient nutrients to carry out vital functions. Maintenance requirements (M) are twice basal energy requirements and *Table 6.7* gives the respective values for rabbits of different bodyweights.

Table 6.7 Daily energy requirements for maintenance

Bodyweight (kg)	Basal (kcal)	Maintenance (kcal)	Approx converted TDN values (g)	Converted EDF values (kcal)
1.70	80	160	50	225
2.04	100	200	60	251
2.95	140	280	78	310
3.96	180	360	100	371

For an adult rabbit weighing 3 kg and consuming 113 g feed per day the metabolisable energy level of the ration would need to be in the region of 2464 kcal per kg (10.309 MJ per kg) to provide a maintenance diet.

Reproduction According to Sandford (1957) the energy requirement for reproduction is between 1M and 1.3M in the first 21 days of pregnancy increasing to 2M in the last 10 days when fetal development is at its greatest (Hammond, 1936).

In practice one diet may be fed at a controlled amount for 21 days and at twice this level for the final period of gestation, or alternatively a diet higher in energy can be fed in late gestation.

From a practical standpoint the doe must not become fat during pregnancy as this will affect reproduction. On the other hand, a doe inadequately fed will be undernourished to the detriment of her milk production.

According to the NRC (1966) the TDN for gestating does should be 58% which is 3% higher than their recommendations for maintenance plus some small growth, as opposed to fattening. The NRC, however, do not make an allowance for early or late pregnancy and the assumption must be therefore that their figure relates to the first 21 days of gestation. Using the figure of 2M for late pregnancy and relating this to TDN produces a recommendation of about 70% TDN for the final period of gestation. This figure compares favourably with the NRC (1966) recommendation of 72% TDN (12.652 MJ per kg).

Lactation The milk yield of does depends both on hereditary factors and on the amount and quality of food given. Peak milk yield is usually reached between the 12th and 28th day of lactation after which it gradually declines until by the 45th day the doe is almost dry. At the height of lactation a 4.5 kg New Zealand White doe will produce between 160 g and 200 g milk per day. A yield of 200 g per day over a six-week lactation period is equivalent to the daily 10–11 gallon yield of a dairy cow.

On the assumption that a rabbit converts food energy into milk at the same rate as the cow, namely about 45%, then with a peak yield of 180 g per day and an energy value at 250 kcal per 100 g, the EDF of the ration needs to be about 1000 kcal. This figure compares favourably with the calculations of Sandford (1957) who cites 560 kcal for a 4 kg doe yielding 112 g milk per day. There is a need to include an estimate for the litter's consumption of some solid food, and reference to the maintenance needs will show that a 4 kg doe needs some 1440 kcal plus a small litter allowance. Thus, the total needs of the doe during lactation may increase to 4M. In terms of TDN, 4M gives a figure of 400 g TDN per day. Relative to a feed intake of 500 g per day this represents a TDN of 80% (14.058 MJ per kg).

Growth Using the data set out in *Table 6.8* and applying the formulae for calculating starch equivalent (SE) (Schurch, 1949) it is possible to arrive at a practical TDN value for the different stages of growth (*Table 6.9*). The conversion from SE to TDN is based on the work by Blaxter (1950) where 0.184 lb SE is equivalent to 0.23 lb TDN.

Table 6.8 Food consumption and bodyweight guide for New Zealand White rabbits

Age (weeks)	Weekly feed intake (kg) Doe and litter	Per young rabbit	Bodyweight Total litter (kg)	Per young rabbit (g)
Kindling to:				
1st week	1.91	–	0.45	56.7
2nd week	2.29	–	1.09	136.2
3rd week	2.31	–	1.91	239.0
4th week	3.10	0.91	4.40	550.0
5th week	5.24	2.32	7.08	885.0
6th week	7.00	4.35	9.80	1226.0
7th week	8.02	5.02	12.35	1544.0
8th week	9.31	6.31	15.61	1952.2

Table 6.9 ME and TDN requirement for growth of New Zealand White rabbits

Age (weeks)	4	5	6	7	8
Approx weight (kg)	0.55	0.88	1.22	1.54	1.95
Weekly weight gain (g)	28.5	48.0	48.7	45.4	57.4
Daily TDN requirement per g gain (g)	34.0	54.0	63.5	75.4	78.5
% TDN in ration	85	85	80	71	63
MJ per kg feed	14.934	14.934	14.058	12.475	11.069

More detailed information of the energy requirements for growth and fattening necessitates knowledge of the ratio with which the energy of the food used for weight gain relates to the energy of the product. The older the rabbit the higher the proportion of fat to protein, and therefore the greater is the energy required per unit of gain.

THE REQUIREMENT FOR PROTEIN

A few years ago the knowledge of protein quality and quantity was extremely vague. Today, although there is still much to clarify, the greater information available has enabled more precise estimates of requirements to be made. As with other farm livestock, the protein requirement is greatest in the early growing period and for the first 20–24 days after birth is adequately met by the doe's milk. After this time the young rabbit is less dependent on the milk and more dependent on the ration fed to the doe. The protein requirement of the pregnant doe is smaller in the early stage of gestation and increases towards the end.

102 *The nutrition of rabbits*

Maintenance

According to Brody (1935) a level of 10% protein is needed for maintenance. The NRC (1966) recommend 12% protein whilst Sandford (1957) lists the daily needs of rabbits varying in weight between 0.9 and 4.9 kg (*Table 6.10*).

Table 6.10 Protein requirement for maintenance of bodyweight

Bodyweight (kg)	0.9	1.3	1.8	2.2	2.7	3.1	3.6	4.0	4.5	4.9
Digestible protein (g per day)	4.1	5.7	7.3	8.4	9.2	10.3	11.1	12.1	13.0	14.1

Pregnancy

During the last 10 days of gestation the demand for protein is greater than in the preceding 20 days. According to Sandford (1957) the protein need in the first 20–22 days of pregnancy is 1.3M, rising to 2M for the final 10 days. For a doe weighing 3 kg the ration would therefore need to contain approximately 16% crude protein.

Lactation

Large amounts of protein are required by the lactating doe and unless this need is met milk production is likely to suffer. A doe suckling an average litter of seven provides an amount of milk relative to the quality of the protein fed. Rabbits' milk is three to four times as rich in protein and fat, half as rich in sugar and almost four times richer in minerals than cows' milk. The figures in *Table 6.11* compare the milk from three species.

Table 6.11 Composition of rabbits', cows' and goats' milk (%) (Morris, 1936)

Species	Water	Protein	Fat	Sugar	Ash
Rabbit	69.5	12.0	13.5	2.0	2.5
Cow	87.3	3.4	3.7	4.9	0.7
Goat	86.9	3.8	4.1	4.6	0.8

Approximately one-third of the energy value of a doe's milk is supplied by the protein. Thus, it is essential that the feed have a high protein content of good quality. According to the NRC (1966) a lactating doe with a litter of seven requires 17% protein. This recommendation is for conventional weaning of the young at seven weeks, followed by remating of the doe.

When does are remated some three to four days after kindling and the young rabbits are weaned at four weeks rather than seven, the nutrient demand for lactation is followed immediately by peak nutrient demand by the unborn litter. There is, therefore, under the postpartum breeding system, a continuously high nutrient demand which is mainly influenced by a high or low remate success rate. If it is high then a doe consuming 340 g food per day will obtain from a 17% protein ration a daily protein intake of 58 g. On the basis of the lactation needs being four times greater than maintenance needs, a ration providing 17% total crude protein would be inadequate unless consumption exceeded 340 g per day.

Growth

It is unfortunate that the NRC (1966) provide only recommendations for normal growth between 2 and 4 kg, because rabbits reared for meat are usually slaughtered at 2 to 2.25 kg. *Table 6.12* shows the feed consumed between four and nine weeks of age. In addition to the increased needs with increasing weight, the amount of daily protein for each unit gain in bodyweight also increases. Smith, Donefer and Mathieu (1960) found no advantage in using 19% rather than 14% protein diets for weaned rabbits. Based on the data of Sandford (1957) for the protein requirements for maintenance, and allowing a requirement of twice maintenance for growth, the figures in *Table 6.12* show the daily protein need and suggested protein percentage in the growing period of four to nine weeks. The high

Table 6.12 Suggested daily and percentage crude protein requirement for fast-growing New Zealand White rabbits

Age (weeks)	4	5	6	7	8	9
Food intake (g per d)	37.0	61.5	85.0	101.0	120.0	132.0
Protein intake (g per d)	7.0	8.2	11.0	12.5	15.5	17.0
Crude protein in ration (%)	28.0	18.0	17.5	17.0	16.5	16.5

suggested protein figure in the fourth week is applicable only to rabbits weaned from does' milk at or before this time. The data shows that a ration containing about 17.5% protein fed from weaning to slaughter will support good growth rate.

THE REQUIREMENT FOR AMINO ACIDS

It has been argued that the amino acid balance of a diet is less critical to rabbits than other non-ruminants because of coprophagy. Proto and Gianini (1969) suggest that coprophagy increases the proportions of methionine, threonine and tyrosine in this type of faecal pellet. Viallard and Raynard (1966), using a diet of oats and lucerne, concluded that micro-organisms in the stomach can play a part in nitrogen metabolism by synthesising protein from urea. On the other hand, King (1971) and Cheeke (1972) could find no evidence to support these findings of Viallard and Raynard.

104   The nutrition of rabbits

Davidson and Spreadbury (1975) hypothesised that if amino acid balance is important then diets of the ruminant type should not support good growth. Their work and that of Gaman and Fisher (1970), Cheeke (1971), Adamson and Fisher (1973), Colin, Arkhurst and Lebas (1974), Kennedy and Hershberger (1974), Spreadbury (1974) and Colin (1975) confirmed that amino acid balance is important for the growing rabbit. Adamson and Fisher (1973) established a list of eleven indispensable amino acids for growing rabbits:

Arginine	Leucine	Threonine
Glycine	Lysine	Tryptophan
Histidine	Methionine	Valine
Isoleucine	Phenylalanine	

Using New Zealand White does aged six weeks (Adamson and Fisher, 1973) or aged four to eight weeks (Davidson and Spreadbury, 1975) amino acid requirements for growth were determined (*Table 6.13*).

Table 6.13 Amino acid requirements for growth

Amino Acid as % of diet	Adamson and Fisher	Davidson and Spreadbury
Arginine	1.00	0.70
Glycine	–	0.50
Histidine	0.45	0.30
Isoleucine	0.70	0.60
Leucine	0.90	1.10
Lysine	0.70	0.90
Methionine and cystine	0.60	0.55
Phenylalanine and tyrosine	0.60	1.10
Threonine	0.50	0.60
Tryptophan	0.15	0.20
Valine	0.70	0.70

As useful as this list of dietary amino acids may be, it is of greater importance to determine the first and second limiting amino acids. As the milk of rabbits has a high methionine plus cystine content relative to the cow, sow and ewe, and the rabbit has a well developed hair coat, it is likely that its need for the sulphur amino acids is greater than that of these other species.

Spreadbury and Davidson (1973) showed that increasing methionine and cystine from 0.39% to 0.53% of feed resulted in an increased body-weight gain of almost 20%. At 0.66% methionine and cystine, weight gain increased over the 0.53% level by only 3.5%. It is suggested that methionine may be the first limiting amino acid for growth with lysine possibly the second limiting. Davidson and Spreadbury (1975) suggested that the arginine requirement is less than previously proposed (Cheeke, 1971) and it is probably less than 0.60% of the diet.

Cheeke (1971), studying the variations in bodyweight gain relative to different lysine levels, showed maximum weight gain at 0.93% lysine when the ration contained 0.40% methionine and 1.09% arginine. Adamson and

Fisher (1973) indicated a lysine requirement of between 0.70 and 0.80% of the diet.

As indicated above, Davidson and Spreadbury (1975) suggested that the arginine requirement is less than 0.60%. However, McWard, Nicholson and Poulton (1967), using two rations of 17% and 23% protein respectively, produced weight gains showing a possible requirement varying according to protein level. With 17% protein 0.99% arginine gave maximum gain and efficiency whilst at 23% protein 1.51% arginine was best. Adamson and Fisher (1973) showed that 1.0% arginine was necessary for maximum weight gain. Davidson and Spreadbury (1975) hypothesised that the arginine requirement is lower than hitherto proposed because some of this essential amino acid is synthesised.

Amino acid needs for gestation and lactation

There is a need for research into the protein and amino acid requirements of breeding rabbits. Lebas and Collins (1973) assume that the amino acid needs may be similar to the requirement for growth but, as pointed out by Davidson and Spreadbury (1975), such a conclusion may be wrong because late gestation and early lactation needs for total protein are higher than during the early growth of weaned rabbits.

RABBIT MILK REPLACER DIETS

Unpublished field trial work using a diet formulated to replace the doe's milk offered as a creep feed to rabbits aged 14 days showed that by 56 days the weekly bodyweight gain exceeded that of the control rabbits by 50 to 60%. The overall growth rate was increased by some 25% at marketing and mortality declined by more than 50% (the national average for mortality varies between 25 and 40%).

Whilst undoubtedly there is a need for much more work in this area, that which has been done is most encouraging and should stimulate further research into rabbit milk replacer diets. The high mortality which occurs when young rabbits change from a milk to a solid diet, especially in four to five week weaning systems, can be largely offset by the provision of a ration higher in nutrient density than is commonly fed. Such a ration has also been found to produce a growth rate curve more resembling a straight line than the uneven curve of conventionally fed rabbits.

THE REQUIREMENT FOR MINERALS

According to the NRC (1966) estimated requirements for minerals are available only for manganese, potassium, magnesium and phosphorus. They give the following requirements for growth: manganese 1.0 mg per rabbit per day; magnesium 30–40 mg per 100 g of diet; potassium 0.6% of diet and phosphorus 0.22% of diet.

Calcium

According to Chapin and Smith (1967a), a calcium level of 0.35 to 0.40% in the diet is necessary for maximum bone calcification. The same authors (1967b), in investigating calcium tolerance, found that 5% reduced growth and percentage ash in femurs. There was no detrimental effect at 3.25% or 0.45% calcium. Besancon and Lebas (1969) found that excess calcium is well absorbed but not well retained. They gave a retention coefficient of 27.6% ± 1.9% and a true digestibility of 51.8% ± 2.8%.

Phosphorus

Chapin and Smith (1967a) found no difference in bone ash, weight gain or feed efficiency with 0.96% P or 1.47% P in the diet. However, increasing calcium from 0.31% to 1.35%, giving a Ca/P ratio 1:1.1 and 1:4.7, significantly increased weight gain and bone ash. It is apparent that the calcium and phosphorus requirements vary and appear to be similar to that of the growing chicken.

Iron

According to Aitken and King Wilson (1962), rabbits' milk is deficient in iron and by the time of weaning, the young may be anaemic. Both iron and copper are necessary to prevent anaemia and the amounts needed to prevent deficiency symptoms are well satisfied by practical rations. Supplementation by 100 ppm iron is recommended and for copper 10 ppm.

Salt

The general recommendation for salt is between 0.5% and 1.0% of the ration. However, the rabbit tolerates much higher levels than this. Gompel, Hammon and Mayer (1936) showed that as much as 35 g salt per 100 ml water could be tolerated if introduced gradually. Most UK rabbit rations contain a salt level between 0.4 and 0.5% but a higher level is necessary to help prevent boredom and chewing. It is suggested that a salt block may be a useful supplement for very intensively housed rabbits.

Table 6.14 summarises the estimated mineral and trace element requirements.

Table 6.14 Summary of estimated mineral and trace element requirements from limited research data

Minerals (% of diet)		Trace Elements (ppm)	
Calcium	0.6 to 1.2	Manganese	50
Phosphorus (total)	0.4 to 0.8	Zinc	20
Salt	0.5 to 1.0	Iron	100
Magnesium	0.25	Copper	10
Potassium	1.0 to 1.7	Cobalt	unknown
		Iodine	unknown

THE REQUIREMENT FOR VITAMINS

Precise requirements are known for only a few vitamins. Roche (1965) provide so called cautious estimates based on available literature. *Table 6.15* shows these estimates as well as estimates for vitamin B_6 and B_{12}.

Table 6.15 Recommended vitamin fortification to compound feedstuffs

Vitamin A[1]	9000 iu per kg	Choline[1]	1300 mg per kg
Vitamin D[1]	900 iu per kg	Vitamin B_6[2]	1 mg per kg
Vitamin E[1]	40 iu per kg	Vitamin B_{12}[3]	10 mg per kg
Nicotinic Acid[1]	50 mg per kg		

[1] Roche, 1965
[2] Hove and Herndon, 1957
[3] Skvorcova, 1963

As a measure of insurance, supplementation by vitamins B_1, B_2, B_6 and pantothenic acid is recommended. The requirement for vitamin C, if any, is completely unknown.

It is known that certain vitamins of the B-complex are synthesised by fermentation in the gastrointestinal tract and their absorption is facilitated by coprophagy. For example, the coprophagic faeces have been found to contain three times as much vitamin B_{12} as normal faecal pellets. Despite this, Skvorcova (1963) found that adding additional vitamin B_{12} to a concentrate ration improved growth rate. According to Hove and Herndon (1957) the rabbit does not synthesise its own vitamin B_6 requirements and they recommend supplementing B_6 at the rate of 1 mg per kg of feed.

THE REQUIREMENT FOR WATER

Rabbits must have free access to water at all times since limiting water access to 6 or 12 hours each day will reduce bodyweight gain. Water allowance should be based on the following data (*Table 6.16*).

Table 6.16 Recommended water allowances per rabbit per day

Non-pregnant does and early pregnant does	0.28 litre
Adult bucks	0.28 litre
Late pregnant does	0.57 litre
Suckling doe, post weaning	0.60 litre
Doe plus litter of seven, aged 6 weeks	2.30 litre
Doe plus litter of seven, aged 8 weeks	4.50 litre

The addition of fresh green foods to the rabbit's diet reduces the requirement for supplementary water; approximately 250 g fresh cabbage each day will theoretically replace the need for supplementary water. However, under all feeding systems a fresh clean water supply should always be available. Withholding water causes a marked reduction in feed intake (Cizek, 1961). An adult rabbit will excrete up to 170 g urine each day (Blount, 1945), and this waste needs to be replaced. In the absence of water the rabbit will consume its own urine, thus increasing the risks of disease. Consumption of water is highest in the evening and lowest in the morning and also greatest if the water is warmed to 40°C (Kalugin and Utkin, 1974).

FEED ADDITIVES

Kurilov *et al.* (1958) found that biomycin improved the weight gain of growing rabbits and that stunted rabbits showed very marked bodyweight increases when biomycin was included in the diet at 1 mg per kg bodyweight. Casady and Hagan (1963) could obtain no growth rate advantage with 50 g zinc bacitracin per ton but claimed a reduction in mortality. King (1974) found an improvement in live weights at two and four weeks of age when 20 g per ton virginiamycin were fed but the differences failed to reach significance.

Whitney (1974) reported that dimetridazole administered via the water at 0.025% appears to be a safe, effective drug for the control of non-specific enteritis. Growth response to modern approved antibiotic and non-antibiotic drugs as used in the pig and poultry industry has been poorly established. What little information there is suggests a small and insignificant biological response.

Based on the available data given it is now possible to provide estimates of requirements for many of the major ingredients of feedstuffs for the different classes of rabbit, and these are set out in *Table 6.17*. Clearly as a more precise and better understanding of these and other requirements develops this list will have to be modified.

Table 6.17 Suggested levels of nutrients for rabbits

Classification	Adult rabbit; non-pregnant does; early pregnant does	Late pregnant does; lactating does and litter	Growing/ fattening rabbits
NUTRIENT			
Protein (%)	12–16	17–18	17–18
Energy (TDN)	65	70–80	80
Energy (MJ/kg)	11.420	12.301–14.058	14.058
Fat (%)	2–4	2–6	2–6
Fibre (%)	12–14	10–12	10–12
Calcium (%)	1.0	1.0–1.2	1.0–1.2
Phosphorus (%)	0.40	0.40–0.80	0.40–0.80
Salt (%)	0.50	0.65	0.65
Magnesium (%)	0.25	0.25	0.25
Potassium (%)	1.0	1.5	1.5
Manganese (ppm)	30	50	50
Zinc (ppm)	20	30	30
Iron (ppm)	100	100	100
Copper (ppm)	10	10	10
AMINO ACIDS (% of diet)			
Methionine plus cystine	0.50	0.56	0.56
Lysine	0.60	0.80	0.80
Arginine	0.60	0.80	0.80
VITAMINS			
Vitamin A (iu per kg)	8000	9000	9000
Vitamin D (iu per kg)	1000	1000	1000
Vitamin E (iu per kg)	20	40	40
Vitamin K (mg per kg)	1.0	1.0	1.0
Nicotinic acid (mg per kg)	30	50	50
Choline (mg per kg)	1300	1300	1300
B_{12} (mg per kg)	0.01	10	10
B_6 (mg per kg)	1.0	1.0	1.0

References

Adamson, I. and Fisher, H. (1973). 'Amino acid requirement of the growing rabbit.' *J. Nutr.*, **103**, 1306

Aitken, F.C. and King Wilson, W. (1962). *Rabbit Feeding for Meat and Fur*. Commonwealth Agricultural Bureaux, Farnham Royal, Bucks.

Arrington, L.R., Platt, J.K. and Franke, D.E. (1974). 'Fat utilisation by rabbits.' *J. Anim. Sci.*, **38**, 75

Axelsson, J. (1949). Standards for Nutritional Requirements of Domestic Animals in the Scandinavian Countries. Ve Congres, Internat. Zootec, Paris

Battaglini, M.B. (1968). 'Importance of coprophagy in domestic rabbits in relation to utilisation of some nutrients.' *Riv. Zootee Agricu. Vet.*, **6**, 21

Besancon, P. and Lebas, F. (1969). 'True digestibility and retention of calcium by growing rabbits given a rich diet in calcium and phosphorus.' *Ann. Zootech.*, **18**, 37

Besedina, G.G.O. (1969). 'Effect of fibre on digestibility of nutrients by rabbits.' *Krolik. Zver*, **4**, 19

Blaxter, K.L. (1950). 'Energy feeding standards for dairy cattle.' *Nutr. Abst. Rev.*, **20**, 1

Blount, W.P. (1945). *Rabbits' Ailments.* Bradford; Fur and Feather

Brody, S. (1935). The Relationship between Feeding Standards and Basal Metabolism, p. 12. Rep. of Conference on Energy Metabolism. Penn. USA

Casady, R.B. and Gildow, E.M. (1959). 'Rabbit nutrition.' *Proc. Anim. Care Panel*, **9** (1)

Casady, R.B. and Hagan, K.W. (1963). *Rabbit Raiser*, Feb., p.8

Casady, R.B., Sawin, P.B. and Van Dam, J. (1971). *Commercial Rabbit Raising*. Agric. Handbook No. 309. Washington, USA; USDA

Chapin, R.E. and Smith, S.E. (1967a). 'Calcium requirements of growing rabbits.' *J. Anim. Sci.*, **26**, 67

Chapin, R.E. and Smith, S.E. (1967b). 'The calcium tolerance of growing and reproducing rabbits.' *J. Anim. Sci.*, **26**, 905

Cheeke, P.R. (1971). 'Arginine, lysine and methionine needs of the growing rabbit.' *Nutr. Rep. Int.*, **3**, 123

Cheeke, P.R. (1972). 'Nutrient requirements of the rabbit.' *Feedingstuffs*, **44**, 28

Cizek, L.J. (1961). 'Relationship between food and water ingestion in the rabbit.' *Am. J. Physiol.*, **201**, 557

Colin, M. (1975). 'Effect of adding lysine to diets based on sesame oilmeal for rabbits.' *Nutr. Abstr. Rev.*, **45**, Abstr. 2714

Colin, M., Arkhurst, G. and Lebas, F. (1974). 'Effects of addition of methionine to the diet on growth of the rabbit.' *Nutr. Abstr. Rev.*, **44**, 7581

Davidson, J. and Spreadbury, D. (1975). 'Nutrition of the New Zealand White rabbit.' *Proc. Nutr. Soc.*, **34**, 75

Ferrando, R., Wolter, R., Vitat, J.C. and Megard, J.P. (1970). 'Teneurs en acides aminés des deux categories de fèces du lapin: caecotrophes et fèces dures.' *C.R. Acad. Sci.*, Serie D., **270**, 2202

Gamon, E. and Fisher, H. (1970). 'The essentiality of arginine, lysine and methionine for the growing rabbit.' *Nutr. Reps. Inter.*, **1**, 57

Gompel, M., Hammon, F. and Mayer, A. (1936). 'Effect of sodium chloride content of the ration on the feeding habits of the domesticated rabbit.' *Ann. Physiol. Physicochim. biol.*, **12**, 504

Hammond, J. (1936). 'Pregnancy in the rabbit.' Weltgeflugel kongress, Berlin, Leipzig, **i**, 153

Hove, E.L. and Herndon, J.F. (1957). 'Vitamin B_6 deficiency in rabbits.' *J. Nutr.*, **61**, 127

Jacquot, R., Le Bars, H. and Simmonnet, H. (1958). *Nutrition Animale. Vol. 1: Données Générales sur la Nutrition et l'Alimentation.* Paris; Baillière

Kalugin, Y.A. and Utkin, L.G. (1974). 'Water requirements of rabbits.' *Nutr. Abstr. Rev.*, **44**, 10, Abstr. 6877

Kennedy, L.G. and Hershberger, T.V. (1974). 'Protein quality for the non-ruminant herbivore.' *J. Anim. Sci.*, **39**, 506

King, J.O.L. (1971). 'Urea as a protein supplement for growing rabbits.' *Br. Vet. J.*, **127**, 523

King, J.O.L. (1974). 'Effects of feeding virginiamycine on the fertility of rabbit does and the development of young rabbits.' *Vet. Rec.*, **94**, 290

Kurilov, N.V., Popov and Novikov (1958). *Vest. sel'.-khoz. Navki, Mosk.*, **10**, 56

Lebas, F. and Colin, M. (1973). 'Effet de l'addition d'urée à un régime pauvre en proteines chez le lapin en croissance.' *Ann. Zootech.*, **22**, 111

McWard, G.W., Nicholson, L.B. and Poulton, B.R. (1967). *J. Nutr.*, **92**, 118

Morris, S. (1936). 'The protein requirements of lactation.' *Nutr. Abstr. Rev.*, **6**, 273

Myers, K. (1955). 'Coprophagy in the European rabbit in Australia.' *J. Austr. Zool.*, **3**, 336

National Research Council (1954). Publs. Natn. Res. Coun., Wash., No. 331

National Research Council (1966). Publs. Natn. Res. Coun., Wash., No. 1194

Parkin, R.J. (1975). Personal communication

Portsmouth, J.I. (1962). *Commercial Rabbit Meat Production.* London; Illiffe

Proto, V. and Gianini, L. (1969). Produz. Anim., 8203

Roche (1965). *Feeding Rabbits.* London; Roche Products Ltd

Sandford, J.C. (1957). *The Domestic Rabbit.* London; Crosby Lackwood

Schürch, A. (1949). 'The theoretical background of rabbit feeding.' *Schweiz. landw. Mh.*, **17**, 27

Skvorcova, I.L. (1963). 'Effect of vitamin B_{12} combined with biomycin on the growth of young rabbits.' *Krolikov. Zverov.*, **9**, 10

Sinquin, P.J. (1973). *Rabbit Production and Marketing.* France; ITAV

Smith, S.E., Donefer, E. and Mathieu, L.G. (1960). 'Protein for growing fattening rabbits.' *Feed Age*, **10**, 7

Spreadbury, D. (1974). 'Protein and amino acid requirement of the growing meat rabbit.' *Proc. Nutr. Soc.*, **33** (2), 56A

Spreadbury, D. and Davidson, J. (1973). *Proc. Nutr. Soc.*, **33** (2), 56A

Templeton, G.S. (1955). *Domestic Rabbit.* Illinois; Interstate Printers and Publishers

Thacker, E.J. (1956). 'The dietary fat level in the nutrition of the rabbit.' *J. Nutr.*, **58**, 243

Viallard, V. and Raynaud, P. (1966). 'Recherches sur l'utilization de l'urée par les microrganisms de l'estomac du lapin.' *C.R. Soc. Biol.*, **160**, 2478

Whitney, J.C. (1974). 'Treatment of enteric disease in the rabbit.' *Vet. Rec.*, **94**, 533

Yoshida, T., Pleasants, J.R., Reddy, B.S. and Wostmann, B.S. (1968). Efficiency of digestion in germ free and conventional rabbits.' *Br. J. Nutr.*, **22**, 723

7
CALCIUM REQUIREMENTS IN RELATION TO MILK FEVER

D.W. PICKARD
*Department of Animal Physiology and Nutrition,
University of Leeds*

Milk fever must have been a considerable problem to dairy farmers for many years. In 1819, Knowlson gave a precise description of the symptoms and was aware of the effect of age and yield on the incidence of the disease, which must have been prevalent in those days in order to warrant the comment 'It is a heavy disorder, and kills many.'

The relationship between milk fever and hypocalcaemia was established by Little and Wright (1925) and Dryerre and Greig (1928) and it is now generally accepted that parturient hypocalcaemia results from the rapid withdrawal of calcium from the blood by the udder at the start of lactation. When the hypocalcaemia becomes severe (usually less than 2.5 mEq l^{-1}) cows show symptoms of milk fever.

The influence of diet on the incidence of milk fever has been the subject of much research and considerable controversy has surrounded the subject of 'steaming up' in relation to milk fever. In the discussion which followed the paper of Dryerre and Greig (1928) it appears that a majority of the participants believed that feeding concentrates before calving resulted in a greater incidence of milk fever. The final word rested with Professor Dryerre who, referring to the divergent views, said that there were 'concentrates and concentrates. Unless one knows exactly the composition of these, the amount and frequency of administration and the weight and condition of the animals, it would be unwise to venture any opinion as to their efficacy.' It now appears that it is the composition of such rations, particularly the content of calcium and phosphorus, and the amount consumed before calving, which governs their effect on the incidence of milk fever. Milk fever commonly occurs where large amounts of dairy rations are fed to dry cows in the last weeks before calving. It can also occur on farms where no supplementary feed is offered prior to calving. These two situations may provide very different intakes of calcium and phosphorus but both produce a similar result – that is, a high incidence of milk fever. This apparent anomaly can be explained in the light of evidence which has recently emerged on the ability of animals to adapt to changes in their intake of calcium.

114 Calcium requirements in relation to milk fever

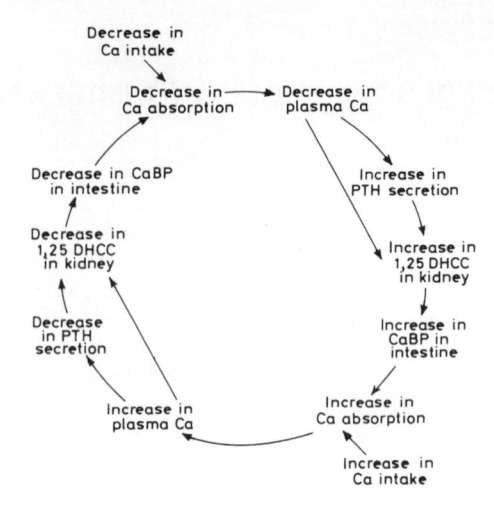

Figure 7.1 Calcium adaptation

When an animal is fed a diet containing a greater amount of calcium than it requires, the proportion of the dietary calcium which is actually absorbed declines. Conversely, an animal adapts to a decrease in the amount of calcium in its diet by absorbing more of it. Calcium adaptation (*Figure 7.1*) has been recognised for some time in rats (Ellis and Mitchell, 1933), cattle (Comar *et al.*, 1953) and humans (Malm, 1963). It is the mechanism whereby an animal adapts to a change in calcium intake which has recently been elucidated.

When an animal such as a pig, or a chick, is fed on a diet low in calcium, changes take place in the metabolism of vitamin D, such that more 1,25 dihydroxycholecalciferol (1,25 DHCC) is formed in the kidney (Boyle, Gray and DeLuca, 1971) from 25 hydroxycholecalciferol, produced in the liver from vitamin D itself. The low calcium diet also results in an increase in the levels of circulating parathyroid hormone (PTH) and the evidence to support the theory that PTH also stimulates 1,25 DHCC production is considerable (Tanaka and DeLuca, 1973). The active metabolite of vitamin D under normal circumstances is 1,25 DHCC (Lawson *et al.*, 1971) and this stimulates the formation of a specific calcium-binding protein in the small intestinal mucosa (Corradino, 1973). This specific calcium-binding protein is involved in the active transport of calcium across the intestinal mucosa (Wasserman and Taylor, 1963), and enables the adapted animal to absorb a greater proportion of its dietary intake of calcium.

When the intake of calcium is increased these mechanisms of adaptation operate in reverse; the secretion of PTH declines and the kidney, instead of producing 1,25 DHCC, produces the 24,25 dihydroxy metabolite of vitamin D. In the absence of 1,25 DHCC the amount of calcium-binding protein in the intestinal mucosa declines and the absorption of calcium is reduced. It must be noted that since calcium adaptation involves changes in the intestinal mucosa, it takes days rather than hours for an animal to adapt to a change in its calcium intake. Animals are also able to adapt to changes in their requirement for calcium; when this increases (as at the beginning of lactation) the animal adapts by absorbing a greater proportion of its dietary intake. It appears that animals attempt to keep the amount of calcium actually absorbed very close to their requirement for calcium irrespective of the amount present in the diet.

Cows which are fed large amounts of dairy rations in the weeks before calving are receiving more calcium than they require because dairy rations contain sufficient calcium to meet the requirements of lactating animals. They adapt to this excess of dietary calcium by absorbing less of it. Since it takes several days for the intestine to change in response to a change in the demand for calcium, the intestine is unable to absorb sufficient calcium to meet the increased demand when calcium is withdrawn from the blood by the udder to produce milk in the day or two before calving. This rapid withdrawal of calcium by the udder results in milk fever. Over-feeding calcium in the dry period also has a detrimental effect on the cow's ability to mobilise calcium from her bones to maintain blood calcium above the danger level at calving.

The calcium intake of cows which are given no supplementary feed before calving may be much closer to their actual requirement at this time and they would absorb a high proportion of the calcium which is eaten. These animals are likely to suffer from milk fever because there is insufficient calcium in the diet to meet the demands for extra calcium just before calving. If a cow produced 15 kg of milk in her udder on the day before calving, this would contain more than 20 g of calcium. Cows on grass with an estimated intake of 50 g of calcium per day would be unable to absorb sufficient calcium to supply the udder at this time. Younger cows may be able to mobilise calcium from their bones and so avoid milk fever, but the availability of calcium from the bones declines with advancing age and this makes older animals more susceptible to milk fever.

The method developed at Leeds for preventing milk fever relies on matching the intake of calcium and phosphorus to the requirement for these elements around parturition. Cows which need supplementary feeding before calving are given a ration low in calcium and phosphorus such as rolled barley or a cereal based compound without minerals, in addition to their normal diet of grass, hay or silage. Normal dairy rations are gradually substituted for the low calcium and phosphorus ration beginning seven days before calving such that cows are receiving a minimum of 4 kg dairy ration by the time they calve.

For animals which are given no supplementary rations before calving, milk fever may be prevented by introducing dairy rations two or three days before calving to coincide with the formation of milk in the udder. When the appetite of the cow is not able to cope with sufficient normal dairy ration, a special ration with a higher calcium and phosphorus content may be needed.

In general, the aim is for an intake of calcium of around 50 g per day and of phosphorus around 30 g per day during the dry period. The

116 *Calcium requirements in relation to milk fever*

intake of both elements should be increased by 50 g per day by the day before calving (Pickard, 1975; Pickard *et al.*, 1975).

The requirement of the dairy cow for calcium in the last month of pregnancy is approximately 15 g per day (Agricultural Research Council, 1965) and using the ARC figure for the availability of calcium, which is 45%, an intake of something less than 40 g per day would be adequate for the cow at this time. The figure of 50 g quoted above is based on this ARC figure and on the fact that under farm conditions an intake of less than 50 g per day would be difficult to achieve (the average composition of herbage being 0.50% DM, and assuming a DM intake of 10 kg per day). It appears that although the figure of 50 g calcium per day is higher than the ARC recommendation, it is not much too high. This may be because the availability of calcium is not a constant 45%. As the ARC acknowledged, the availability of calcium falls as the intake increases.

Figure 7.2 shows that the availability of calcium falls in a linear manner as intake increases, but it appears more likely that the relationship is curvilinear, with the result that the net absorption of calcium might remain constant over a wide range of calcium intakes. There are at present insufficient data available to prove this point conclusively but it is suggested that it is only at intakes of calcium above 50 g per day that the suppression of parathyroid gland activity, the reduction in 1,25 DHCC production and the depressed efficiency of calcium absorption become severe enough to cause embarrassment to the cow, as she adapts to a situation where her net requirement for calcium has increased by a factor of 2 or 3 over a period of hours around the time of parturition.

Whilst the ARC were compelled, for the sake of simplicity in presenting tables of requirements, to take a figure of 45% for the availability of calcium, they were careful to draw attention to the fact that

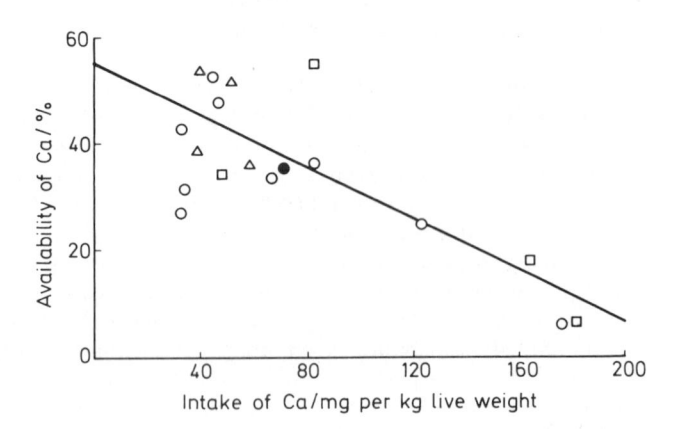

Figure 7.2 The relationship between calcium intake and its availability in cattle weighing 300–400 kg. ○ = *values for individual heifers and steers;* ● = *mean for Hereford cattle;* △ = *means for steers;* □ = *values for individual cows (ARC, 1965)*

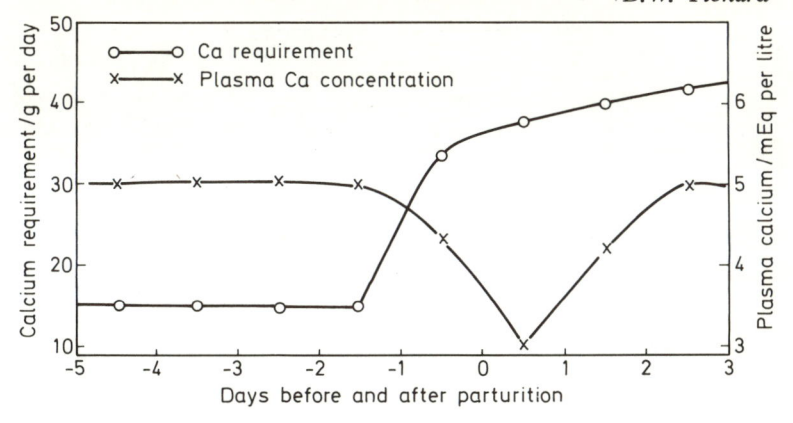

Figure 7.3 Diagram of changes in calcium requirement and in plasma calcium concentration at parturition

calcium availability does indeed fall as intake increases. This point is very important when dealing with the dry cow which has a low absolute requirement for calcium and a very high potential for its consumption.

Recent experience indicates that where dry cows are fed according to the ARC requirement for calcium, the incidence of milk fever is low. A refinement is necessary, however, because the calcium requirements of the cow increase *before* parturition, not afterwards. Milk is usually formed in the udder before calving and the fall in plasma calcium levels begins one or two days before calving.

Figure 7.3 illustrates in diagrammatic form the changes which take place in the level of calcium in the plasma, coincident with changes in the absolute requirement for calcium. It is therefore advisable to increase the intake of calcium before parturition.

The involvement of phosphorus in milk fever is important and will be dealt with next. Comparisons may be made between the high phosphorus diet recommended for dry cows by Boda and Cole (1954) and the relatively low calcium and phosphorus diet advocated here. It is certainly true that a high phosphorus diet will suppress calcium absorption and stimulate the parathyroid glands, and in many cases a reduction in the incidence of milk fever has been achieved. It does not work in all cases and there are two possible reasons for this. Firstly, since the requirement for calcium increases before calving, the level of phosphorus in the diet should be reduced to allow more calcium to be absorbed. Secondly, it is clear that a high phosphorus diet interferes with the metabolism of vitamin D by inhibiting the kidney 1-hydroxylase (MacIntyre, 1975). Without 1,25 DHCC, calcium absorption and bone mobilisation are likely to remain unstimulated.

It was shown by Black *et al.* (1973) and Pickard *et al.* (1975) that a fairly low calcium and phosphorus diet will allow the parathyroid glands to respond adequately to a fall in plasma calcium at parturition. In other words, there is no need for a high phosphorus diet in order to achieve active parathyroid glands.

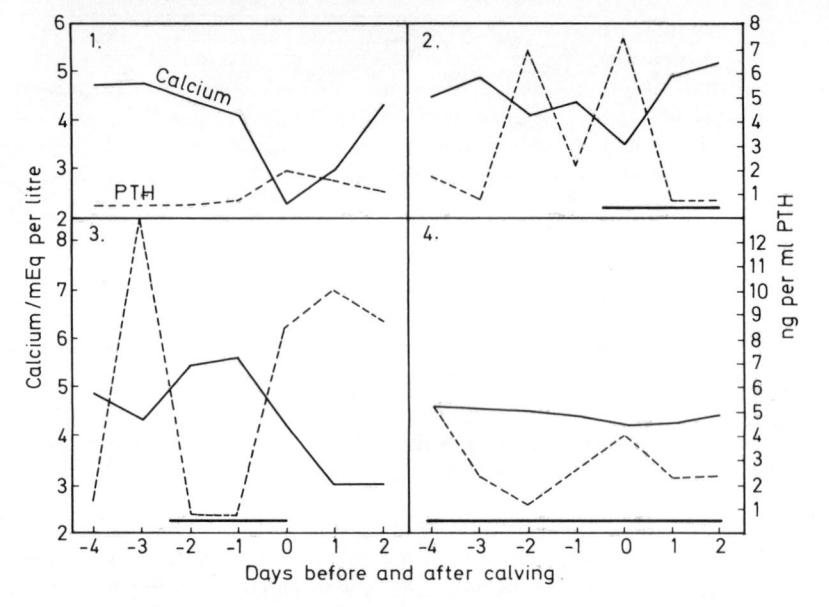

Figure 7.4 Plasma calcium and PTH levels in calving cows. 1. This cow was given a steaming-up ration which supplied an additional 20 g Ca and 15 g P from 4 to 5 weeks before calving. The PTH response was poor, despite the very low plasma calcium level. 2. This cow received no supplementary ration until the time indicated by the solid bar when 50 g Ca and 50 g P were supplied per day. There was a good PTH response which showed a clear inverse relationship with plasma calcium. 3. This cow was treated in a similar manner to No. 2 but the calcium and phosphorus supplement (50 g per day of each) was fed only for the time indicated by the solid bar. Failure to maintain calcium and phosphorus intake after calving resulted in a decline in the plasma calcium level. Again there was a good PTH response. 4. This cow was fed the supplementary 50 g Ca and P per day for 6 days before calving – calving date is not always easy to estimate. The parathyroid glands were still able to respond at calving, indicating that it takes several days for adaptation to the higher calcium intake to be completed

Figure 7.4 shows the results from four cows. This indicates that with an intake of calcium of approximately 50 g and phosphorus 30 g per day, the parathyroid glands do not appear to have been suppressed, and that by increasing the intake of calcium and phosphorus before calving it is possible to prevent the plasma calcium level from falling to the low levels which result in milk fever. Low phosphorus intakes are occasionally stated as the cause of milk fever but it appears that although phosphorus deficiency is not uncommon, the intake of calcium is often too high, and this may reduce the availability of the phosphorus. Much has been spoken of the importance of the calcium to phosphorus ratio in relation to milk fever. The work of Gardner in the last few years (Gardner, 1970; Gardner and Park, 1973) indicated that a ratio of 2.3:1 was optimal for the prevention of milk fever but Beitz, Burkhart and Jacobson (1974) could not agree with this conclusion. When referring to ratios, the actual amounts of elements should be borne in mind. A ratio of calcium to phosphorus of 2:1 would have a very different effect on the parathyroid and vitamin D status of a dry cow when fed at the rate of 100 g calcium per day than the same ratio at a calcium intake of 50 g per day. The correct amount of calcium and phosphorus fed to match a cow's requirements will always give the correct ratio. A correct ratio does not necessarily give the correct amounts.

Since in most species the ability of the skeleton to mobilise calcium declines with age, it is the intestine which must take over the role of supplier of the extra calcium needed by the older cow at the beginning of lactation. When the dry cow is fed on a fairly low intake of calcium and phosphorus, the intestine adapts by increasing the efficiency of absorption of calcium from the intestine. Increasing the intake of calcium and phosphorus just before parturition will allow more calcium to be absorbed at the time when plasma calcium levels are beginning to fall. The beneficial effects of this system of feeding are illustrated in *Figure 7.5*. The plasma calcium levels were maintained at a significantly higher level in the treated group of cows than in those steamed up in a traditional manner (Pickard *et al.*, 1975). Rasmussen (1973) has indicated that when the plasma calcium level falls below 2.5 mEq l^{-1} the ability of PTH to mobilise calcium from the bones is greatly reduced. By increasing the dietary intake just before calving it appears that it is possible to maintain the plasma calcium well above this dangerously low level. The importance of continuing to feed high levels of calcium after calving must be stressed. Most cases of milk fever occur within 48 hours of calving and every effort should be made to maintain the cow's appetite over this period.

Field trials have recently been carried out from Leeds in an attempt to put some of this theory into practice, and some results are now available.

Figure 7.6 includes results from the 24 farms that have returned results so far. A total of 216 cows was selected from these farms.

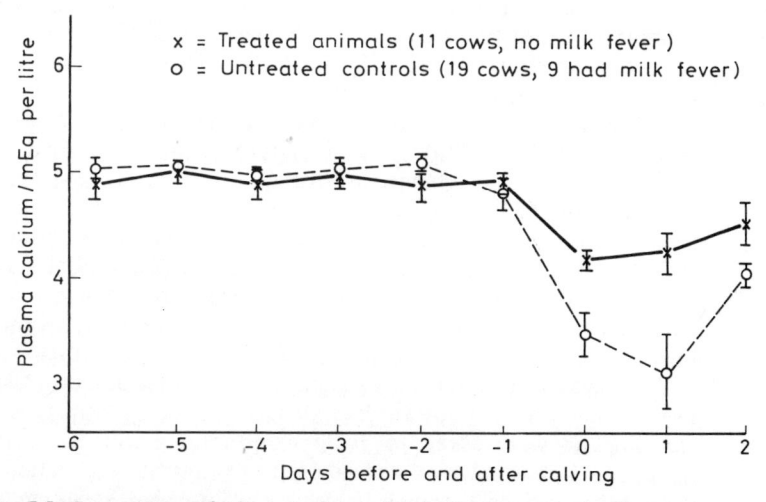

Figure 7.5 Comparison of plasma calcium levels in two groups of cows, one treated with Ca and P, the other untreated

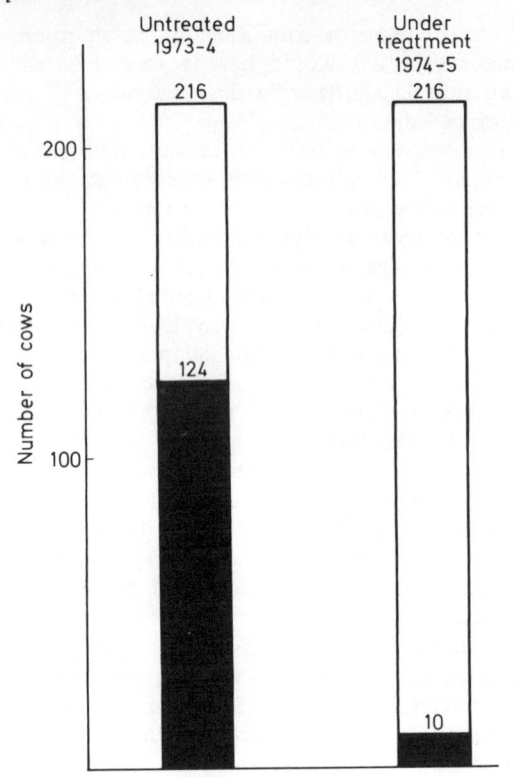

120   *Calcium requirements in relation to milk fever*

Figure 7.6 Effect of regulation of Ca and P intake on incidence of milk fever in farm trials

They were the high risk cases, either because they had had milk fever in 1973–74 or were considered by their owners to be very likely to go down this time. In fact, 124 of the cows had previously had milk fever but only 10 cases occurred in 1974–75. Seven of these 10 cases were on farms with a history of hypomagnesaemia and it is known that PTH action is impaired when the magnesium status is low (Estep *et al.,* 1969; Muldowny *et al.,* 1970). The hypomagnesaemia may have contributed to the partial failure of the treatment on these farms. On this point it should be noted that high magnesium intakes are also likely to cause trouble because of the known inhibitory effect of high levels of magnesium on the absorption of calcium from the intestine. Despite the fact that these results only allow a comparison to be made between different years, the data came in throughout the year, and indicate that the incidence of milk fever is related to the intake of calcium and phosphorus.

How was the intake of calcium and phosphorus controlled on these farms with a variety of feeding systems operating? In general, those steaming up were asked to avoid dairy rations and either feed rolled barley or a low calcium and phosphorus steaming-up ration, and to introduce dairy rations just before calving. Those not steaming up were asked to increase the intake of calcium and phosphorus before calving by introducing dairy rations or a special high calcium and phosphorus ration which supplied sufficient extra calcium and phosphorus when fed at the rate of 2 kg per day.

More results have been received recently, and there are very few reports of failures from farms where the steaming-up system of feeding has been followed. The practice of increasing the calcium intake just before calving without steaming up has failed on three farms, where it was subsequently found that the calcium content of the herbage was higher than average, and the content of phosphorus tended to be low. Manston (1967) found that when the intake of calcium was increased before calving, on an already high calcium intake, the incidence of milk fever increased. It is obviously more difficult to guess the calcium and phosphorus intake where steaming up is not practised and care should therefore be taken to assess the likely intake of calcium and phosphorus from the herbage of such farms. Where farmers are unwilling to substitute their high calcium grass with low calcium concentrates, milk fever must be controlled by other means.

In conclusion, it appears from this work that where dairy cows are fed levels of minerals, especially calcium and phosphorus, which are as close as possible to their requirements for these minerals around the time of parturition, the incidence of milk fever is likely to be much reduced.

Summary

It is believed that the two major causes of milk fever are:

1. Overfeeding calcium and phosphorus by steaming up with dairy rations.
2. Underfeeding calcium and phosphorus by failing to increase the intake of calcium and phosphorus just before calving when the requirement for these elements suddenly increases.

Milk fever may be prevented by:

1. Avoiding steaming up with dairy rations and using a low calcium phosphorus ration instead.
2. Increasing the calcium and phosphorus intake by 50 g per day for each element, beginning 2–3 days before calving.

References

Agricultural Research Council (1965). *Nutrient Requirements of Farm Livestock, No. 2, Ruminants.* London; Agricultural Research Council

Beitz, D.C., Burkhart, D.J. and Jacobson, N.L. (1974). *J. Dairy Sci.*, **57**, 49

Black, H.E., Capen, C.C. and Arnaud, C.D. (1973). *Lab. Invest.*, **29**, 173

Boda, J.M. and Cole, H.H. (1954). *J. Dairy Sci.*, **37**, 360
Boyle, I.T., Gray, R.W. and DeLuca, H.F. (1971). *Proc. natn. Acad. Sci. USA*, **68**, 2131
Comar, C.L., Monroe, R.A., Visek, W.J. and Hansard, S.L. (1953). *J. Nutr.*, **50**, 459
Corradino, R.A. (1973). *Science*, **179**, 402
Dryerre, H. and Greig, R. (1928). *Vet. Rec.*, **8**, 721
Ellis, M. and Mitchell, H.H. (1933). *Am. J. Physiol.*, **104**, 1
Estep, H., Shaw, W.A., Wathington, C., Hobe, R., Holland, W. and Tucker, St. G. (1969). *J. clin. Endocrin.*, **29**, 842
Gardner, R.W. (1970). *J. Dairy Sci.*, **53**, 682
Gardner, R.W. and Park, R.L. (1973). *J. Dairy Sci.*, **56**, 385
Knowlson, J.C. (1819). *The Cattle Doctor*. London and Otley; Wm. Walker & Sons
Lawson, D.E.M., Fraser, D.R., Kodicek, E., Morris, H.R. and Williams, D.H. (1971). *Nature*, **230**, 228
Little, W.L. and Wright, N.C. (1925). *Br. J. exp. Path.*, **6**, 129
MacIntyre, I. (1975). In *Calcified Tissue*. Edited by S. Pors-Nielson and E. Hjørting Hansen. Copenhagen; FADC Publishing Co
Malm, O.J. (1963). In *The Transfer of Calcium and Strontium across Biological Membranes*. Edited by R. Wasserman. New York; Academic Press
Manston, R. (1967). *J. agric. Sci.*, **68**, 263
Muldowny, F.P., McKenna, T.J., Kyle, L.H., Freaney, R. and Swan, M. (1970). *New Engl. J. Med.*, **282**, 61
Pickard, D.W. (1975). *Br. vet. J.*, **131**, 744
Pickard, D.W., Care, A.D., Tomlinson, S. and O'Riordan, J.L.H. (1975). *J. Endocrin.*, **67**, 45P
Rasmussen, H. (1973). *Triangle*, **12**, 103
Tanaka, Y. and DeLuca, H.F. (1973). *Archs Biochem. Biophys.*, **154**, 566
Wasserman, R.H. and Taylor, A.N. (1963). *Nature*, **198**, 30

8

PROTEIN QUANTITY AND QUALITY FOR THE UK DAIRY COW

W.H. BROSTER and
J.D. OLDHAM
*National Institute for Research
in Dairying, Reading*

Introduction

The general philosophy of judging an allowance of protein for a dairy cow is based on converting her output of nitrogen, as controlled by her size and her milk output, to dietary supply in proportion to efficiency of utilisation. The traditional approach has been feeding experiments from which are judged the particular amounts of protein above which no further improvement in yield is obtained. This is then declared to be the required rate of feeding. Supplementary evidence from nitrogen balance trials has been obtained in some experiments and has provided estimates of digestible protein and of biological value, i.e. efficiency of utilisation of digested protein. Efficiency of utilisation has hitherto been based on digestibility as the first stage and then efficiency of utilisation of digested protein as the second stage. The digestible fraction of dietary nitrogenous compounds has been variously described as digestible crude protein, digestible true protein, protein equivalent. All these units and biological value itself have been undermined as knowledge of utilisation of dietary protein expands, encompassing gradually the appreciation of the use of non-protein nitrogen (NPN). Evidence is now sufficiently strong in this area that attempts can be made to formulate protein requirements from efficiencies of the various stages of nitrogen metabolism. In this paper two questions are therefore posed: 'What does the cow do with the protein supply — how is output affected?' and 'How efficient is metabolism of protein?'. Both questions seek information leading to statements of amounts of protein to feed. Feeding trials dominate the first question; metabolism studies are providing the key answers to the second. For both approaches the major factor emerging from recent research is the role of energy supply in protein utilisation. This will be highlighted in this review.

124 *Protein quantity and quality for the UK dairy cow*

Perspective in Assessing Protein Requirements

Almost universally the experimental approach adopted has been to assess protein utilisation over short periods and at energy intakes equal to requirements. This is a logical but limited starting point. Forced on the research worker to a considerable extent by shortage of experimental facilities it precludes assessment of a number of long-term issues: adaptation to a diet (Jackson, 1974); early lactation mobilisation of protein as well as fat from the body in the establishment of peak yield (Paquay, de Baere and Lousse, 1972); later recouping of such body losses; effect of protein on fertility and general health (Broster, 1972; Hewitt, 1975); effect of protein intake on persistency of milk yield (Thomas, 1971; Broster and Bines, 1974). All these long-term considerations have been seriously neglected. Requirements of protein for pregnancy (*Table 8.1*) have been documented (Jakobsen, 1957, quoted by Agricultural Research Council (ARC), 1965). It is necessary to recall these long-term issues to maintain perspective in considering protein requirements for an animal from which long-term as well as current high output is sought.

Two further issues allied to protein requirements may also be conveniently dealt with here. Firstly it is generally acknowledged that nitrogen intake can affect consumption of forages (Balch and Campling, 1962). The level of protein consumption at which this occurs to an appreciable

Table 8.1 Protein requirement of cows during pregnancy (45 kg calf) (Jakobsen, 1957, interpreted by ARC, 1965)

Months of pregnancy	N retained (g per day)	Available protein (g per day)
5–6	1.7	15
7	5.1	45
8	12.0	110
9	29.0	260

Table 8.2 Responses to additional groundnut cake in the concentrates fed to dairy cows receiving a basal ration of silage (Murdoch, 1962)

Treatment	A	B	C	D	s.e. mean
Silage intake* (kg per day)	36.55	36.18	39.91	41.09	±0.86
Concentrates (kg per day)	5.58	5.58	5.58	5.58	–
% groundnut in concentrates	0	8	16	25	–
Milk yield (kg per day)	15.75	16.57	17.34	17.25	±0.03
SNF (%)	8.90	8.84	8.87	8.78	±0.05
Non-protein N in milk (mg N per 100 ml milk)	19.3	21.5	22.6	25.5	–

*Silage of 11% digestible crude protein in the dry matter

Table 8.3 The effect of level of intake on apparent digestibility of protein (%) in three isocaloric diets (Broster et al., 1973—4)

Diet (% concentrates:% hay)		60:40	75:25	90:10
Digestible energy intake	3.0	73.38	74.60	75.28
(Multiples of maintenance)	3.8	72.67	74.27	75.76
	4.3	72.45	76.39	76.90

extent, less than 10% of the dry matter, is below that reasonably anticipated in diets for lactating cows. However Ørskov et al. (1971) observed responses to additional protein in lambs up to 16—20% protein in the dry matter; Broster, Tuck and Balch (1964) reported appreciable effects from groundnut meal on hay intake of heifers in late pregnancy — a lactation phase when intake may be critically low; and Murdoch (1962) reported beneficial effects on silage intake and hence on milk yield of additional protein in the concentrates given to lactating cows (*Table 8.2*). Indeed the role of protein in consumption and utilisation of forage is not adequately understood (Campling, 1964; Campling and Murdoch, 1966; Tagari et al., 1965, 1971; Griffiths, Spillane and Bath, 1973; T. Smith, 1976).

Secondly Moe, Reid and Tyrrell (1965) and Wagner and Loosli (1967) have observed large falls in N digestibility to occur as intake of maize-based rations increases. Wiktorsson (1971) and Broster et al. (1973—4) (*Table 8.3*) (*see also* Broster et al. (1971—2)) have not found this with diets based on grass, hay or silage. No firm conclusions can be made on digestibility of nitrogen in this regard.

Feeding Experiments on Protein Requirements

PROTEIN AND ENERGY INTERRELATIONSHIPS

Generally experiments have included energy intakes equal to nominal requirements with variation in protein supply (Broster, 1972). This simple situation belies the complexities of the problem that the use made of protein is markedly dependent on the energy supply. The evidence on this is meagre for dairy cows though it is more plentiful for growing cattle and for sheep (Broster, 1972, 1973).

Elliott, Reed and Topps (1964) described the relationship between protein and energy with a quadratic equation. Balch (1967) modified this, retaining its general curvilinear nature (*Figure 8.1*). Black and Griffiths (1975) derived linear functions for the two phases of the response curve to protein intake: for a low intake range with protein limiting growth; and for a higher range of intake with protein no longer limiting growth. The equation by Elliott, Reed and Topps (1964) is of the general form y (growth) $= b_1 C + b_2 P + b_3 P^2 + b_4 CP + k$ [where C is energy intake, P is protein intake, $b_1 \ldots _4, k$ are constants]. This includes an additive and a multiplicative beneficial effect from each nutrient. Since b_4 is universally negative the effects of high protein intake are detrimental. Output from a given level of protein intake is

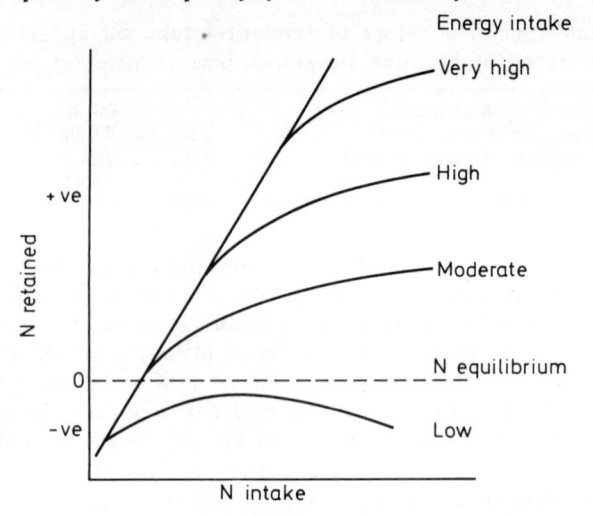

126 Protein quantity and quality for the UK dairy cow

Figure 8.1 Diagrammatic model of the relationship between intake of nitrogen and nitrogen retained in heifers receiving various amounts of dietary energy (after Balch, 1967)

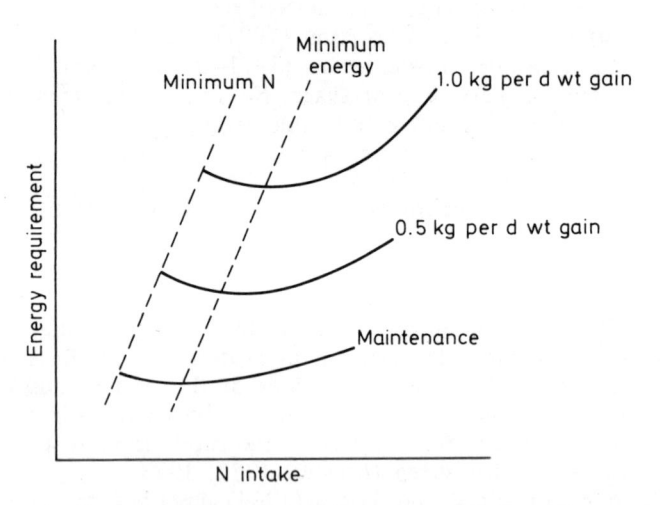

Figure 8.2 Diagrammatic representation of the amounts of digestible protein and digestible energy required for various rates of growth in young cattle (based on Stobo and Roy, 1973)

dependent on energy supply; and, vice versa, amount of protein to sustain a particular output depends on energy supply. These requirements can be set out by a reconstruction of the above relationship (Broster, 1973) (*Figure 8.2*) as for milk N ouput by Robinson and Forbes (1970). Minimum intakes of protein and energy can be distinguished. They do not, it must be pointed out, occur together: minimum protein supply requires more than minimum energy supply.

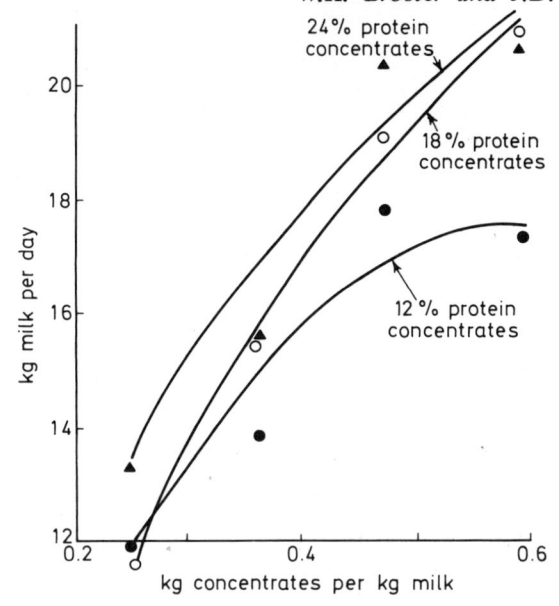

Figure 8.3 Effect of rate of concentrates feeding and % crude protein in the concentrates on milk yield (kg per day) of cows (Gordon and Forbes, 1973, unpublished results)

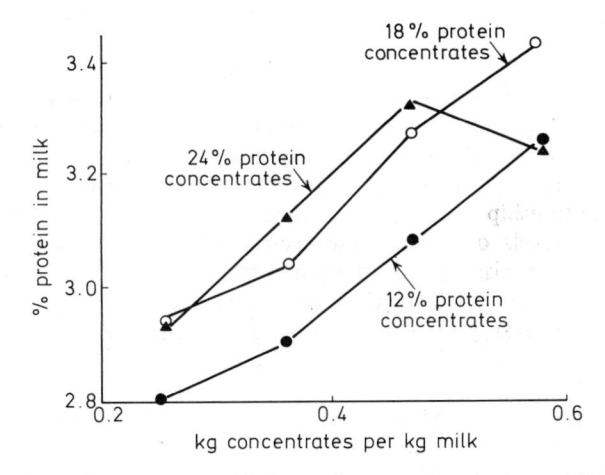

Figure 8.4 Effect of rate of concentrates feeding and % crude protein in the concentrates on milk protein % of cows (Gordon and Forbes, 1973, unpublished results)

Evidence from studies including independent variation of energy and protein within experiments is quite inadequate to construct a corresponding model for dairy cows. The few experiments reported (*see* Broster, 1972) confirm that energy intake does influence the milk output supported by a given amount of protein.

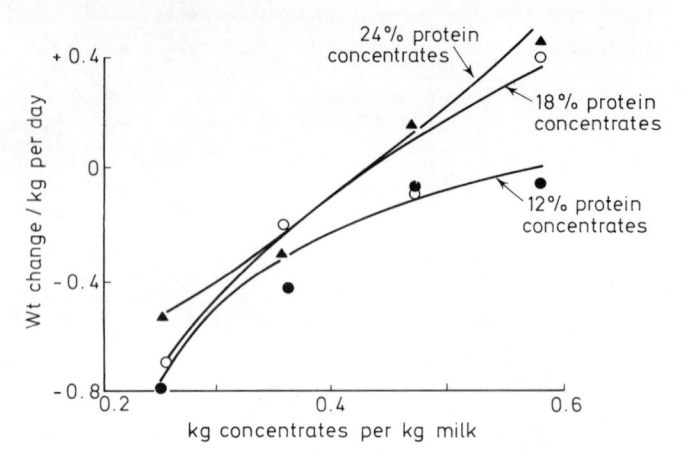

Figure 8.5 Effect of rate of concentrates feeding and % crude protein in the concentrates on live weight change (kg per day) of cows (Gordon and Forbes, 1973, unpublished results)

More recently Gordon and Forbes (1973), quoted by Bines and Broster (1974), have produced the most comprehensive experiment on the issue. It is used here to illustrate the relationships involved. In a randomised block experiment with 72 cows, using a 7 week experimental period, four rates of concentrates feeding were arranged factorially with three levels of protein intake. Concentrates varied from 0.2 to 0.6 kg per kg milk and, for each level, concentrates of 12, 18 and 24% crude protein were used. Curvilinear responses to each nutrient were obtained. *Figures 8.3, 8.4* and *8.5* give the results for milk yield, milk protein content and live weight change. It is immediately clear that, in general, 12% crude protein is insufficient and 24% excessive; also that there is a response relationship between milk output and inputs of energy and protein. At low levels of concentrates levels 18% or 24% protein contents were not as beneficial compared with 12% as at high rates of concentrates consumption. The effect of increasing energy supply was greater than that of increasing protein. Live weight changes will be dealt with on p.20.

OPTIMAL RATIO OF PROTEIN TO ENERGY IN THE DIET

Paquay *et al.* (1973) have estimated optimal dietary ratios of metabolisable energy to protein (*Table 8.4*). Too wide a ratio reduces milk yield; too narrow a ratio is not beneficial. The optimal ratio of protein to energy was judged to fall with increasing time from calving.

The ARC (1965) pointed out that for their estimates of protein requirements, increasing metabolisable energy (ME) density in the diet must be accompanied by increased protein contents. These must increase at a greater rate than the energy density because of greater efficiency of utilisation of ME in diets of high energy concentration.

Table 8.4 Optimal protein–energy ratios in the diets of dairy cows according to stage of lactation (Paquay et al., 1973)

Months of lactation	Optimal ratio of g digestible N per MJ ME
1–3	2.2
4–5*	2.4
6–7	1.7
8–9*	1.7
10+	1.3

*Few values (authors' comment)

REQUIREMENTS VERSUS RESPONSES

Taking then the limited case of cows adequately fed for energy Broster (1972) estimated from the results of 16 experiments the nature of the response curve to variation in protein intake. Over the lower range of intakes the response to additional protein was 0.44±0.058 kg milk per 0.1 kg additional digestible crude protein intake. At high levels of intake the response was 0.06±0.002 kg milk per 0.1 kg additional digestible crude protein intake. Thus at high levels of intake the response is so small that the point of inflexion in the curve can be regarded as the requirement. The predicted requirement was 58.5±1.82 g digestible crude protein per kg milk with confidence limits ($P = 0.05$) of 52.3–60.0 g per kg milk. The unit of intake – digestible crude protein – is challenged later in this chapter (Miller, 1973b quoted by Bines and Broster, 1974).

Some specific examples of the effects of underfeeding protein will draw out the significance of the response concept. A reduction of intake from 100% to 60–70% standards (Woodman, 1957) caused the following falls in milk yield: 1.4 kg in 10.4 kg (Rowland, 1946); 0.9 kg in 11.3 kg (Breirem, 1949); 2.7 kg in 18.1 kg (Rook and Line, 1962). At 80% standards (Woodman, 1957), falls in milk yield of 0.7 kg in 20.4 kg (Frens and Dijkstra, 1959); 0.9–1.8 kg in 18.1 kg (Rook and Line, 1962); 0.6 kg in 18.1 kg (Broster et al., 1969); 0.5 kg in 12.2 kg (Broster et al., 1960) were reported. These falls represent changes in output with change in input and they are as important as the concept of a requirement. The latter is a rigid, oversimplified version of food utilisation. The concept of responses permits the pinpointing of particular combinations of nutrient inputs for a given animal to support a given level of performance.

For comparison the change in milk output with change in energy intake is 0.1 kg milk per MJ ME for a cow yielding 20 kg at ARC (1965) rates of feeding (Broster, 1974).

MILK COMPOSITION

Figure 8.4 shows the unpublished data of Gordon and Forbes (1973). Level of energy in the diet affects protein content of the milk; so does protein content up to 18% protein in the concentrates but not beyond. Quoting extensive studies by Rook and Line (1961), Balch (1972) plotted the relationship between energy intake and milk solids-not-fat content shown in *Figure 8.6*. The effect was due almost wholly to an increase in protein content, increases occurring in all the major milk proteins, casein, β-lactoglobulin and α-lactalbumin. Though Gordon and Forbes showed some effect of protein intake at low levels of consumption on protein content of the milk, in general protein intakes around normal levels have little effect on protein content though extra may increase NPN in the milk (Rook, 1961). Thus in practice it is energy supply rather than protein supply which is of major interest in milk protein content.

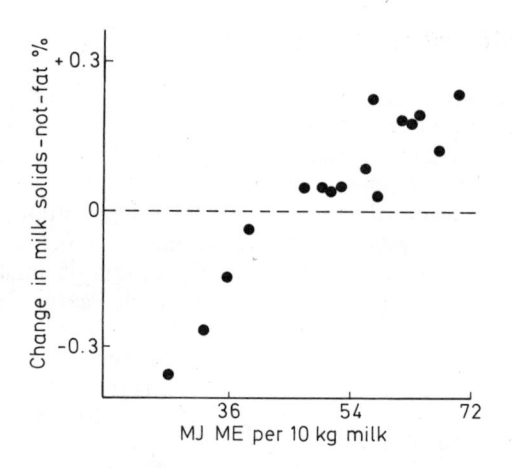

Figure 8.6 Effect of level of energy intake on milk solids-not-fat content (after Balch, 1972)

FACTORIAL ESTIMATES OF PROTEIN REQUIREMENTS

The ARC (1965) used the summation of the nitrogen requirements for various body functions and output pathways as a measure of net N requirements. These in turn were converted into dietary N by application of various factors estimating rates of utilisation. In general these latter values presented minimal requirements. That they undercut some traditional estimates has been noted (Bines and Broster, 1974). They do represent, as the authors comment, estimates of minimal requirements *without* safety margins. As such they are not necessarily directly comparable with values for present feeding standards which are allied to energy intakes equal to requirements and which may include safety margins.

VARIATION IN ESTIMATES OF REQUIREMENTS

Safety margins apart, not all experimental results can be reconciled to the above estimates of protein requirements. Evidence by Drori and Folman (1970) is a notable exception. They found that 125% of the protein level in the milk was sufficient to sustain high yields over whole lactations. Huber (1975) quoted various pieces of evidence that fluctuate in estimates of protein requirements between 12.8 and 17% crude protein and he concluded that 15% is sufficient in early lactation — catering in part for low food consumption at this time — and 12% later in lactation. This draws out the distinction between requirements at different stages of the lactation, building up to the long-term issues referred to earlier.

There is the basic question of optimum percentage protein in the diet for cows of differing yield level. In their experiment, Cuthbert, Thicket and Wilson (1973) used four levels of crude protein in the dry matter (10, 12, 14 and 16%). A family of response curves was obtained for cows of different milk yield capacities. Greater responses were obtained from the higher yielding cows (greater than 20 kg milk yield per day), for which it was concluded 60 g digestible crude protein per kg milk was the required rate of feeding. Less was required for lower yielders. In general inadequate evidence is available on the higher yielding cow that now forms the national herd. Cuthbert, Thicket and Wilson's (1973) figure does not exceed the mean value from published data (Broster, 1972) and *Table 8.8* (p. 148). In these the 19:1 confidence limits were 52–60 g digestible crude protein per kg milk. Van Es' (1972) figure of 0.3 kg digestible crude protein per 500 kg live weight for maintenance has been accepted for this particular calculation. The variation in size of response to change in protein intake was notably small. Even so reference to the classical experiment by Frederiksen *et al.* (1931) (*Figure 8.7*) shows the variation in response

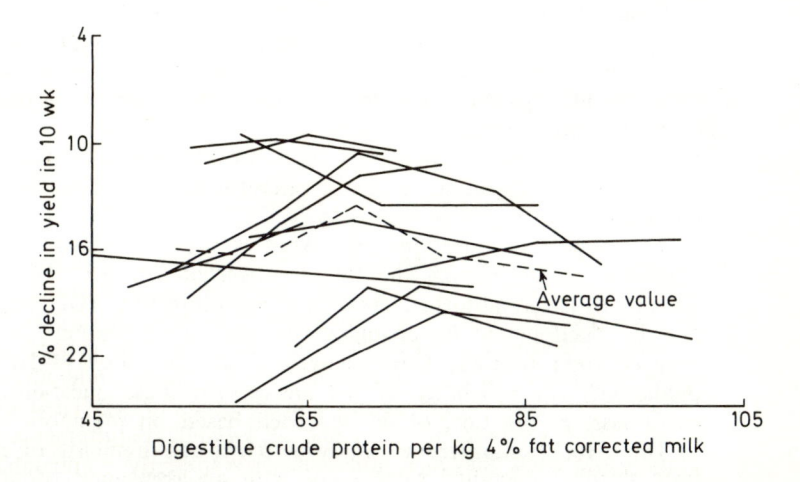

Figure 8.7 Graphical representation of responses in milk production to variation in protein supply (Frederiksen et al., 1931)

that occurs, even within a trial. Local managerial and biological variation is partly the cause of this but also differences of interpretation of units (digestible crude protein, digestible true protein, protein equivalent), of the use of non-protein nitrogen, and of the efficiency of utilisation of absorbed nitrogenous compounds.

Not least of the causative factors has been the inadequate appreciation of the part in protein metabolism played by the energy supply in the rumen and in the body (Oldham, 1973). It is therefore timely and proper to bring into the analysis of the problem the increasing volume of evidence regarding metabolism of protein in the cow, to provide a correction to misapplication of half knowledge and to provide an alternative and more basically orientated approach to estimation of protein requirements.

Efficiency of Metabolism of Protein as a Guide to Requirements

As with energy utilisation (Blaxter, 1967) so with protein, a series of coefficients of efficiency of utilisation at the various key stages of metabolism can provide, when integrated, an estimate of protein requirements. Several workers (e.g. Miller, 1973a; Ørskov, 1976; Satter and Roffler, 1975) have reasoned that sufficient evidence now exists to justify consolidation of the approach into tabulations of amounts of protein to feed. The merits of this line of attack are several: its scientific basis is obvious; it draws protein quality directly into the system; it forms a cross-check on feeding trial estimates of requirements; and it assists understanding of the variation in results encountered there. As with the metabolisable energy system it is capable of absorbing new information into the logic of the approach; the basic pattern caters for all classes of ruminants. This section summarises the approach, indicates the present state of knowledge, and provides an estimate of protein requirements for lactation based on it.

Response to increases in protein input has been shown to depend on energy supply. This indicates an interaction between protein and energy which determines the manner in which protein will be used. Protein requirements, quantity and quality, must therefore be stated in relation to defined energy inputs. This provides a point of reference in studying the transformations which occur to food protein and NPN during fermentation in the rumen, digestion in the abomasum and intestines and utilisation in the host body.

The important questions are: What is the quantitative role of rumen microbes in determining the fate of ingested nitrogen? What happens to protein which is absorbed from the gastrointestinal (GI) tract and can its fate be manipulated by changing the quality of protein absorbed or the type of food offered? How accurately can the value of NPN in feeds be predicted? The reliability of the answers to these questions determines the usefulness, in practice, of an approach based on metabolic considerations.

The ARC Committee to study nutrient requirements of ruminants has been deeply concerned since 1973 with an assessment of protein requirements based on metabolic considerations, and will publish the results of their deliberations shortly (Blaxter, 1976). Kaufmann and Hagemeister

(1975) in W. Germany and Burroughs, Nelson and Mertens (1975) and Satter and Roffler (1975) in the United States have suggested practical approaches. What follows is our assessment of the situation in relation to the UK dairy cow.

RUMEN MICROBIAL GROWTH AND ENERGY SUPPLY

The rumen contains bacteria and protozoa in proportions which depend on a host of factors, many of which are interdependent, for example, source of dietary carbohydrate and rumen pH. It is usually tacitly assumed that the energy cost of rumen microbial growth is a direct reflection of the energy cost of bacterial growth and most published estimates relate only to bacteria. It will, similarly, be assumed here that measurements for bacteria can be applied to the entire rumen population and the term 'energy cost of microbial growth' will be used for all measurements. It must, however, be borne in mind that the presence of a large protozoal population may impose an extra energy burden on the rumen. As protozoa rely to a large extent on bacteria for their protein supply, protozoal protein production includes an extra energy cost for resynthesis of protein. If protozoa merely sequester in the rumen and make only a minor contribution to microbial protein flowing from the rumen, as suggested by Weller and Pilgrim (1974), then a further level of energy inefficiency may be involved.

The relationship between bacterial growth and energy supply is complex (Stouthamer and Bettenhaussen, 1973). It is likely that the most important variable which may affect microbial growth yield in the rumen is the dilution rate of rumen fluid (Harrison *et al.*, 1975). No doubt other variables play a part in the broad range of values found for the energy cost of rumen microbial growth. Many of these values are drawn together in *Table 8.5*. We have chosen to relate microbial N production to the amount of organic matter which is apparently digested in the rumen (rumen ADOM). This differs from the amount of organic matter truly digested (rumen TDOM) by the amount of OM which is re-incorporated into microbial biomass (*Figure 8.8*). Although rumen TDOM is the better measure of total energy made available from the fermentation rumen ADOM represents the excess energy which is required to resynthesise degraded food OM into microbial OM. Microbial growth can therefore be related to rumen ADOM (A $\propto$ B in *Figure 8.8*). In addition rumen ADOM is the quantity which is actually measured whereas rumen TDOM is derived from this by making allowance for microbial OM from a knowledge of microbial protein supply from the rumen. This is a hazardous procedure as microbial OM composition is quite variable (McAllan and Smith, 1974; Czerkawski, 1975).

The mean of the values in *Table 8.5* indicates that 32 g microbial N are synthesised per kg rumen ADOM. This value will be used to cover all situations but we are aware that it is not a biological constant and is associated with a degree of variability. It is a difficult value to use in practice as it does not relate to a readily identifiable fraction in food. It

134 *Protein quantity and quality for the UK dairy cow*

Table 8.5 The relationship between microbial N yield and the apparent digestion of organic matter in the rumen (rumen ADOM). For each source the range of values shown refers to all experimental diets used in that experiment

Source	g Microbial N per kg ADOM (rumen)	Ration constituents
DAIRY COWS		
Hagemeister and Pfeffer (1973)	32–47	Hay + barley, concentrates
Hagemeister and Kaufmann (1974)	20–35	Hay + barley, concentrates
SHEEP		
Hume (1970a, b)	23–33	Purified diets
Hogan and Weston (1970)	22–32	Clover or grasses
Hogan and Weston (1971)	32–55	Alkali-treated straw
Lindsay and Hogan (1972)	37–61	Lucerne hay or dried clove
Leibholz (1972)	11–21	Various
Ørskov, Fraser and McDonald (1972)	29–35	Barley + urea
Miller (1973a)	34–40	No details
Hume and Purser (1975)	28–26	Clovers
Sutton *et al.* (1975)	27–38	Hay + barley, concentrates
Overall mean	32	

is possible, though, to predict microbial N yield from the energy content of the food. Miller (1973a, b) related microbial N production to metabolisable energy (ME) intake. This is consistent with the ARC (1965) approach to defining energy allowances and will also be adopted here. There is generally a strong relationship between digestible organic matter (DOM) intake and rumen ADOM (*Figure 8.9*). Most rations conform to one pattern, the major exception being ground and pelleted forage diets. Apart from these, rumen ADOM represents 65% of DOM (*Figure 8.9*). It should be noted that Tamminga (1975) has recently reported values for dairy cows falling in the range 47–54% for the proportion of digestible organic matter apparently digested in the rumen. Nevertheless the consensus value of 65% will be used here. If DOM has an energy content of 18.5 MJ kg^{-1} and ME is 82% of DE (Blaxter, 1967) then microbial N production is:

$$\frac{32 \times 0.65}{18.5 \times 0.82} = 1.371 \text{ g microbial N per MJ ME}$$

This defines the maximum quantity of microbial nitrogen which can be synthesised for a fixed ME intake. The quantity and quality of food nitrogen needed to supply this amount of N is governed by:

1. The efficiency of conversion of food N to microbial N.
2. The degradation of food N in the rumen.

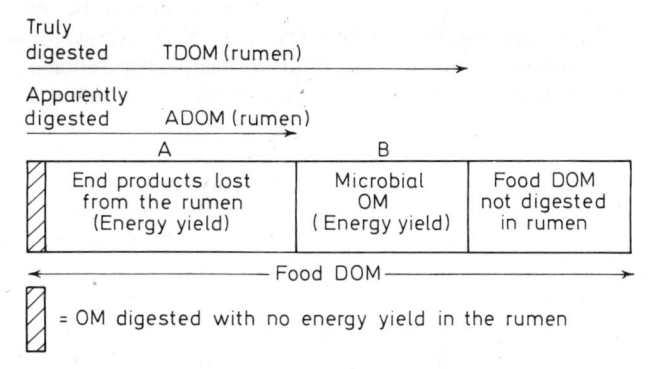

```
Truly
digested    TDOM (rumen)
           ─────────────→
Apparently
digested    ADOM (rumen)
           ──────────→
      A            B
┌──┬──────────┬──────────┬──────────┐
│▨▨│End products│ Microbial│ Food DOM │
│▨▨│ lost from  │    OM    │   not    │
│▨▨│ the rumen  │ (Energy  │ digested │
│▨▨│(Energy yield)│ yield) │ in rumen │
└──┴──────────┴──────────┴──────────┘
         ←──────── Food DOM ────────→

▨▨ = OM digested with no energy yield in the rumen
```

Figure 8.8 The partition of digested organic matter showing the relationship between true (TDOM) or apparent (ADOM) digestion of organic matter in the rumen. See text for further explanation

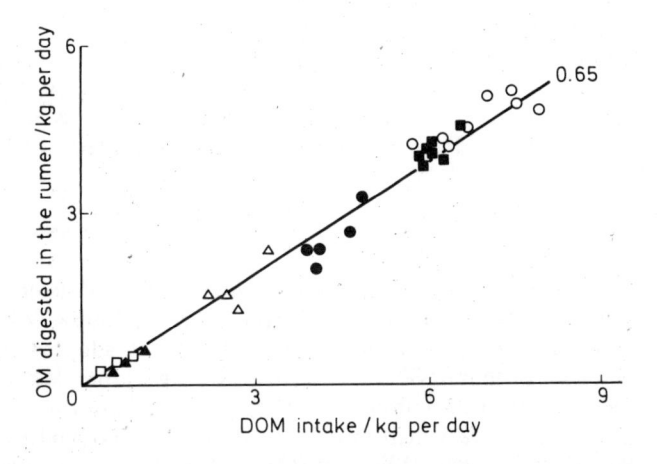

Figure 8.9 The proportion of food digestible organic matter (DOM intake) which is apparently digested in the rumen. ○ *Klooster and Rogers (1969);* ● *Pfeffer, Kaufmann and Dirksen (1972);* ▲ *Beever et al. (1972);* △ *McGilliard (1961);* □ *Nicholson and Sutton (1969);* ■ *Calculated from Hagemeister and Kaufmann (1974). The line at 0.65 digestion in the rumen is shown*

There is a shortage of information on both points. Studies with ^{15}N in sheep show that a large part (35–78%) of bacterial N is derived from ammonia, but the efficiency of capture of rumen ammonia into microbial protein is diet-dependent (Pilgrim et al., 1970; Mathison and Milligan, 1971; Nolan and Leng, 1972): 80–85% of rumen ammonia may be incorporated into microbial protein with highly fermentable diets (fresh lucerne, rolled barley), but only 45–55% with hays. The efficiency of capture of ammonia is therefore high when energy supplies are adequate. As maximum microbial protein yield is achieved at low ammonia concentrations (Satter and Slyter, 1974; Buttery, 1976) inevitable losses of rumen ammonia by rumen fluid outflow or by absorption should be small when the correct balance between fermented OM and N is achieved. When recycled urea-nitrogen is taken into account the apparent efficiency of capture of ammonia from food N approaches 100%. This refers to ammonia only. The microbial N which is not derived from ammonia may include amino acids or peptides of food origin which have been directly incoporated into microbial protein. These are probably incoporated with high efficiency as the free amino acid content of rumen fluid is low (Mangan, 1972). Consequently we have assumed that degraded food N is converted to microbial N with an apparent efficiency of 100%.

There is insufficient information to justify the use of individual degradability values for each food and an umbrella value is an undesirably broad generalisation. We have chosen to place foods in groups according to their degradation characteristics (Table 8.6). The experimental basis for these tentative estimates is unsatisfactory — greater definition is urgently required.

The above information makes it possible to determine, at a given ME intake, the amount of food protein needed to maximise microbial N production or the quantity of urea which can be used to replace food protein. In isolation, however, it reveals nothing of the animal's requirement for protein. To gain knowledge of this it is necessary to examine the course of digestion and utilisation of protein after it leaves the rumen.

QUANTITATIVE PROTEIN SUPPLY TO THE RUMINANT

Protein N represents 72–88% of microbial N (Weller, 1957; Purser and Buechler, 1966; McAllan and Smith, 1972; Burris et al., 1974; Smith and McAllan, 1974). A median value of 80% is adopted here. Microbial protein N yield is thus 0.80 × 1.371 = 1.097 g N per MJ ME intake. This predicts the microbial protein N supply to the absorptive area of the gut. The protein mixture which enters the duodenum normally consists of at least 50% microbial protein. Other proteins presented for digestion and absorption derive from food protein not degraded in the rumen, and endogenous secretions. The proportion of protein N absorbed between the proximal duodenum and the terminal ileum is not influenced greatly by the ratio

Table 8.6 The proportion of food protein degraded in the rumen (Compiled by Dr. R.H. Smith (NIRD))

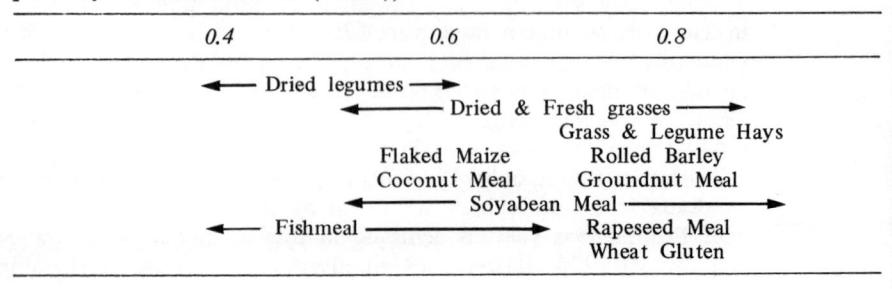

0.4	0.6	0.8
← Dried legumes →		
	← Dried & Fresh grasses →	
		Grass & Legume Hays
	Flaked Maize	Rolled Barley
	Coconut Meal	Groundnut Meal
	← Soyabean Meal →	
← Fishmeal →		Rapeseed Meal
		Wheat Gluten

Table 8.7 The apparent absorption of protein N* from the small intestines of ruminants. For each source the range of values shown refers to all experimental diets used in that experiment

Source	Apparent absorption coefficient	Ration constituents
CATTLE		
Klooster and Rogers (1969)	0.61–0.71	Semi-purified concentrates
Sharma, Ingalls and Parker (1974)	0.65–0.72	Hay + semi-purified concentrates and grass
Watson, Savage and Armstrong (1972)	0.52–0.69	Dried grass + concentrates
Tamminga (1975)	0.69–0.77	Dried grass or grass silages
SHEEP		
Clarke, Ellinger and Phillipson (1966)	0.45–0.76	Hay + maize and soya supplements
Ørskov et al. (1971)	0.51–0.71	Rolled barley + urea or fishmeal
Coelho da Silva et al. (1972a)	0.67–0.79	Dried grass, chopped or pelleted
Coelho da Silva et al. (1972b)	0.66–0.71	Lucerne, chopped, cobbed, or pelleted
MacRae et al. (1972)	0.62–0.64	Dried grass + protein supplements
Ørskov, Fraser and McDonald (1972)	0.64–0.68	Rolled barley + urea
Hogan (1973)	0.61	Clover
Ørskov and Fraser (1973)	0.31–0.81	Barley + soyabean meal
MacRae and Ulyatt (1974)	0.65–0.75	Grasses
Ørskov et al. (1974)	0.61–0.69	Barley + fishmeal and urea

*Where possible values refer to uptake of amino acid N between the proximal duodenum and terminal ileum but where necessary values for uptake of non-ammonia N or total N are included

of microbial:undegraded food protein even when the food protein is of a type which is highly digestible in non-ruminant species. Available data variously refer to the disappearance of non-ammonia N, amino acid N or total N in the small intestine. Much of this information is drawn together in *Table 8.7*. From this it is apparent that most situations are adequately described by allowing a value of 0.7 for the proportion of abomasal or duodenal N absorbed in the small intestine. There is at present no justification for using separate values for microbial and food protein.

PREDICTION OF PROTEIN N SUPPLY

To predict protein N supply and absorption for a given ration the following steps are taken:

1. Define ME intake (I). Hence maximum microbial protein N supply = $1.097 \times I = MP$.
2. Define food protein N intake and calculate degradable food protein N (DP) and undegradable food protein N (FP).

3. If DP ⩾ microbial N calculated as MP/0.8 (to account for protein N content of microbial total N) then microbial protein N production is maximal. If theoretical microbial N > DP the difference in N is the quantity of urea N which should be added to the ration to maximise MP.
4. Total protein N supply to duodenum = MP + FP.
5. Absorbed protein N = 0.7 (MP + FP).

These calculations allow prediction of protein supply to the tissues. The response of the tissues to this protein input is then vitally important in determining protein quantity and quality requirements.

THE UTILISATION OF ABSORBED PROTEIN

Protein which has been absorbed from the gut is used to maintain body tissues and to synthesise new body protein and milk. The efficiency with which these processes are achieved is dependent on the quantity and amino acid composition of the protein and on the supply of energy.

The definition of efficiency used here is the proportion of absorbed protein which is deposited in body protein and milk. It is strictly an input/output relationship and is clearly related to the old established term, biological value (BV). In fact, the efficiency of utilisation of protein made available for absorption is BV in the truest sense; but it is the BV of protein which leaves the abomasum not the BV of food protein, and to use the same term here would introduce a considerable degree of confusion. The intervention of rumen microbial fermentation renders BV, as defined by the ARC (1965), redundant for ruminants. The same can be said for digestible crude protein (DCP). The value of apparently digested protein depends on the conversion of protein digested in the rumen to microbial protein. DCP on its own does not give a clear indication of the processes concerned with crude protein digestion. An alternative to BV is required which relates to the protein which is presented for absorption. The most useful term is the efficiency of utilisation of absorbed protein.

The efficiency of utilisation of protein for milk production is the main concern here but the efficiency of utilisation of protein for body maintenance must not be forgotten. The net utilisation of absorbed protein contains components for both maintenance and milk production. In considering the utilisation of absorbed protein we are really considering the response to a factor which limits the usefulness of the protein. The limiting factor may be an individual amino acid, a group of amino acids or some other factor such as the content of α-amino nitrogen. Success in identifying the limiting factor for milk production has been very limited. Clark (1975) has reviewed much of the recent work in which milking cows have been supplemented, post-ruminally, with amino acids, glucose or casein hydrolysate. Repeated efforts to define a first limiting amino acid for milk production have proved inconsistent. Methionine or phenylalanine may be the most likely choices but Clark (1975) noted that the greatest response in milk production has been produced by duodenal supplementation with casein. Increases of 10–15% in milk protein yield have been achieved. The active factor in casein remains unidentified. The amino acid composition of casein is ideally suited to boost milk production but it is unlikely that duodenally supplemented protein will reach the udder with its amino acid composition unchanged. Differential digestibilities of individual amino acids in the small intestine (Clarke, Ellinger and Phillipson, 1966; Coelho da Silva et al., 1972a, b) and differential use of amino acids for gluconeogenesis will inevitably alter the amino acid composition of supplementary protein which reaches the udder (Black et al., 1968).

The involvement of amino acids in gluconeogenesis may be important in determining the efficiency with which absorbed protein is used. As Clark (1975) suggests, increases in milk protein yield following post-ruminal supplementation with glucose may be the result of sparing amino acids otherwise required for gluconeogenesis. Evans et al. (1975) fed high or low roughage rations to milking cows and observed an increased milk protein yield and increased plasma glucose utilisation rate with cows on the low roughage rations. They suggested that insulin may have had a mediating influence, being stimulated by an increase in glucose supply and itself stimulating protein synthesis and inhibiting gluconeogenesis from amino acids. The net result, that increasing glucose supply rate allowed increased milk protein output, is important. The source of increased glucose supply may be twofold, either arising from increased rumen propionate production or increased duodenal starch passage with high concentrate diets. Sutton (1976) found the available data on rumen volatile fatty acids (VFA) production in cows inconclusive. High concentrate diets (8–13% hay) may produce greater amounts of propionate (mmol per kg digestible dry matter intake) than low concentrate diets (35–55% hay) but there are too few reliable measurements to define the situation properly. On the other hand up to 300–400 g starch can be digested daily in the small intestine of cattle (Sutton, 1976), a significant part of the daily requirement for glucose, which has been estimated by Armstrong and Prescott (1971) to be 1494 g per day for a 590 kg cow producing 20 kg milk per day. Manipulation of the carbohydrate component of the diet may therefore be a means of influencing amino acid utilisation in the body provided that, by increasing glucose supply to the tissues, amino acids are indeed released from gluconeogenic requirements to milk protein production. This would only apply to glucogenic amino acids, thus influencing the composition of amino acids available for milk synthesis as well as the total quantities of amino acids.

The information is not available which would allow these concepts to be incorporated into a description of the efficiency with which absorbed protein is utilised. It is apparent, though, that the 'quality' of absorbed protein may be important in determining responses in milk protein output and that there may be important interactions between absorbed carbohydrate and protein which determine, to some extent, the routes of utilisation of absorbed protein (i.e. gluconeogenesis or milk protein). There will also be an interaction between absorbed energy and protein which may be expected to comply with the diminishing returns law.

In each of these areas there is a woeful lack of data. While there is a considerable body of evidence to describe the course of protein digestion up to the point at which protein is absorbed from the gut of the ruminant, the course of its subsequent utilisation remains impossible to quantify at present. Purser (1970) noted this most powerfully and we can but call again for an increased research effort in this area. For our present purpose, that of defining protein allowances for lactation, it is nevertheless vital that we describe the efficiency of utilisation of absorbed protein. To do this we have chosen a range of values, from 65% to 85%, to compare the estimates of required protein allowances which derive from these with published estimates for minimum required protein allowances. This approach is less than ideal and due care is required in interpreting the results. We emphasise that the comparison is not a means for predicting the correct efficiency of utilisation by comparison with published schemes but is merely an investigative tool to demonstrate the effect which varying efficiencies may have on estimates of requirements.

Maintenance

A value of 59 g CP per day may be used for the daily endogenous urinary N excretion (EUN) of a 500 kg cow (Roy, 1975) but EUN is only part of the protein requirement which is usually termed the maintenance requirement. The other component to be considered is truly endogenous, or metabolic, faecal nitrogen.

Losses of nitrogen from gut wall detritus and endogenous secretions have been described as metabolic faecal nitrogen and assessed as proportional to dry matter intake (ARC, 1965). The quantity involved is assessed indirectly by extrapolation of regressions of faecal N on N intake or by direct measures of faecal N output on low N diets. Such losses form a further demand for dietary N for replacement. The factorial estimation of requirements allows for this, e.g. the adjustment of available protein to digestible crude protein for maintenance of a 500 kg cow is about 80 g added to 115 g for maintenance (ARC, 1965). An approach based on metabolic efficiency of protein utilisation requires cognizance also of these losses, ideally measured as the actual amount of N lost from the gut rather than as empirical estimates of metabolic faecal N (MFN). Such information is not available. Therefore estimates have been made here. Apparent MFN contains a component which is microbial in origin (Mason, 1969). Therefore ARC (1965) may overestimate MFN. The fraction of endogenous N appearing in the faeces will be related to urea recycled to the gut, also to cell fragments from gut tissue, and to endogenous secretion not reabsorbed. Information is lacking on quantitative aspects of these three components. In the absence of definitive evidence an estimate of MFN based on supply of endogenous N arising from turnover of gut protein would fall broadly within the range 25–45 g d^{-1} for a 500 kg cow. Here a value of 35 g d^{-1} has been used in calculations.

Calculation of Protein Requirements

Total protein requirements were calculated as follows. ARC (1965) estimates of ME requirements and DM intakes for a 500 kg cow were used for different levels of yield of milk of 3.35% protein content. These were added to estimated maintenance requirements based on the above discussions. *Figure 8.10* and *Figure 8.11* show estimates of protein requirements with and without endogenous faecal N (EFN).

To compile these figures total requirement for duodenal protein supply was calculated for the chosen efficiencies of utilisation of absorbed protein and compared with potential microbial protein supply at the required ME intake. Where necessary, due allowance was made for undegraded food protein which was required and, by reference to DM intake, the required ration CP % calculated. The procedure is clarified by an example – a cow producing 20 kg milk per day at 80% efficiency of utilisation of absorbed protein.

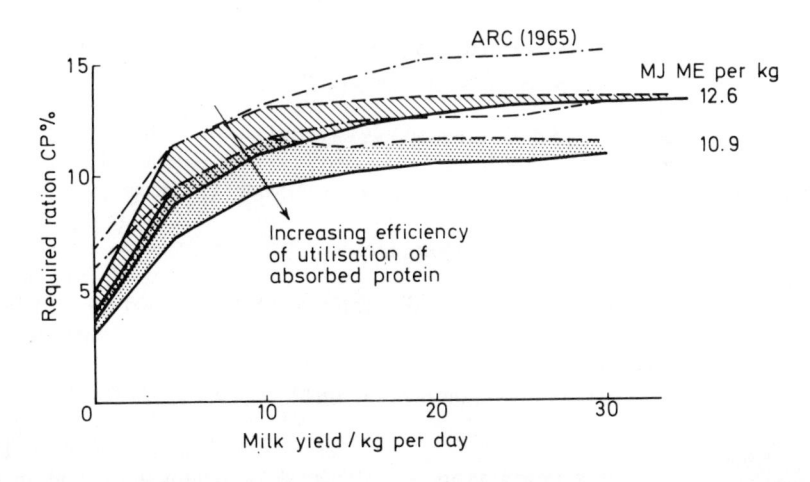

Figure 8.10 Predicted requirements for protein, as % CP in food DM, in relation to milk yield. Requirement was assumed to be milk protein output + EUN. Calculations were made for rations of two M/D values (10.9 and 12.6 MJ ME per kg) and for different efficiencies of utilisation of absorbed protein. The ARC (1965) recommended allowances are shown for both M/D values for comparison (–·–·–). For further explanation see text

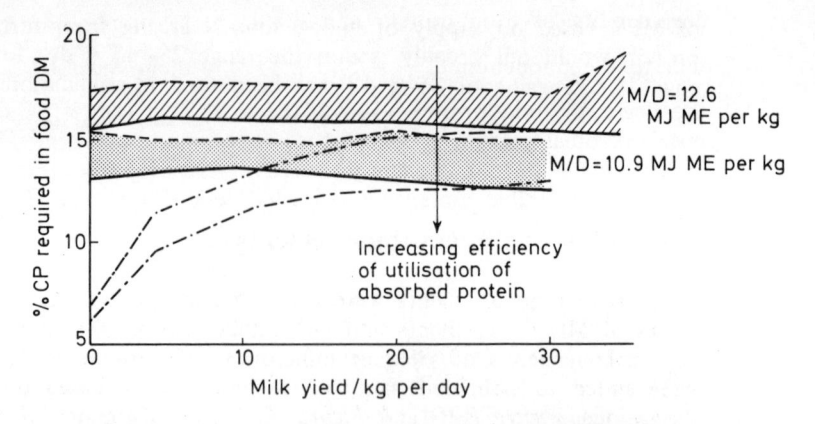

Figure 8.11 Predicted requirements for protein, as % CP in food DM, in relation to milk yield. Requirement was assumed to be milk protein output + EUN + truly endogenous faecal N (at 35 g N per day). Calculations were made for rations of two M/D values (10.9 and 12.6 MJ ME per kg) and for different efficiencies of utilisation of absorbed protein. The ARC (1965) recommended allowances are shown for both M/D values for comparison (—·—·—). For further explanation see text

Protein content of milk = 0.0335 × 20 kg = 670 g

Maintenance protein requirement = 59 g (excluding EFN; this would be 278 g if EFN of 35 g N per day were included)

Total protein output = 729 g per day

At 80% efficiency of utilisation of absorbed protein:

Absorbed protein requirement $= \dfrac{729}{0.8} = 911$ g

This is derived from duodenal protein which is 70% absorbed:

Duodenal protein required $= \dfrac{911}{0.7} = 1302$ g

From ARC (1965), ME requirement for 20 kg milk yield = 156 MJ per day and DM intake for a ration containing 10.9 MJ per kg = 14.3 kg per day

Maximum microbial protein yield = 6.25 × 1.097 (*see* p. 136)

= 6.86 g protein per MJ ME intake

For this ration (156 MJ ME intake) maximum microbial protein yield = 156 × 6.86 = 1070 g per day

This is less than the requirement for duodenal protein (1302 g per day) so the difference (1302 − 1070 = 232 g) must come from undegraded food protein. For this example, taking an overall value of 0.6 for the degradability of food protein in the rumen, to supply 232 g food protein to the duodenum $\frac{232}{0.4}$ = 580 g food protein must be fed. Note that the degraded food protein (580 × 0.6 = 348 g) is insufficient to maximise microbial protein production so that urea could profitably be used as a supplement to make up the deficit. As microbial protein is only 80% of microbial N the amount of degraded food CP to maximise microbial protein = $\frac{1070}{0.8}$ = 1338 g food CP.

Total food CP requirement is made up of:

Degraded food CP (including urea supplement) =	1338 g
	+
Undegraded food CP	= 232 g
Total food CP required	= 1570 g

Therefore 1570 g CP is required in 14.3 kg DM. Required CP % = $\frac{1570}{143}$ = 11%

The results of many such calculations are gathered together in *Figures 8.10* and *8.11*.

The requirements are given as % CP required in DM when DM intake is defined by the ARC (1965). Calculations have been made for rations with two different ME concentrations (M/D) of 10.9 or 12.6 MJ kg^{-1} (2.6 or 3.0 Mcal per kg) and for a range of values for efficiency of utilisation of absorbed protein, from 65 to 85%. The figures show only those values for which the rumen requirement for protein (1.097 g microbial protein per MJ ME) is met but not exceeded, i.e. where there is no wastage of degraded protein in the rumen. The protein requirement is thus met by maximum microbial protein supply plus undegraded food protein supply.

VARIATION IN PREDICTED PROTEIN REQUIREMENTS

Required ration CP % is dependent on the energy concentration of the diet. This is simply because, to meet ME requirements, diets of high ME concentration require lower DM intakes; thus to meet a fixed CP requirement the protein concentration of the diet must rise. If absorbed protein is used with high efficiency, required dietary CP % is reduced. The notched lines in *Figures 8.10* and *8.11* indicate the upper limit to CP % above which, to meet duodenal needs for protein, protein starts to be wasted in the rumen. This happens particularly with highly degradable diets (*see Table 8.6*) and at low efficiencies of utilisation of absorbed protein. The calculations indicate that, dependent on the energy concentration of the diet, and for a required level of milk output, there is a level of ration CP % above which degraded protein is lost from the rumen. When the maintenance requirement is equated with EUN (*Figure 8.10*) the limiting value for ration CP % is 11.5% (M/D = 10.9 MJ kg^{-1}) or 13.3% (M/D = 12.6 MJ kg^{-1}) for milk yields in excess of 10 kg per day. If EFN is taken into account for the maintenance component (*Figure 8.11*) these levels rise to about 15% (M/D = 10.9) and 17.5% (M/D = 12.6) and are constant for all levels of production.

It is of importance to consider the response of cows to incremental changes in diet CP %. Below a limiting value responses are partly a function of increased microbial protein production and partly a function of undegraded food protein reaching the duodenum (0.8 unit increase in

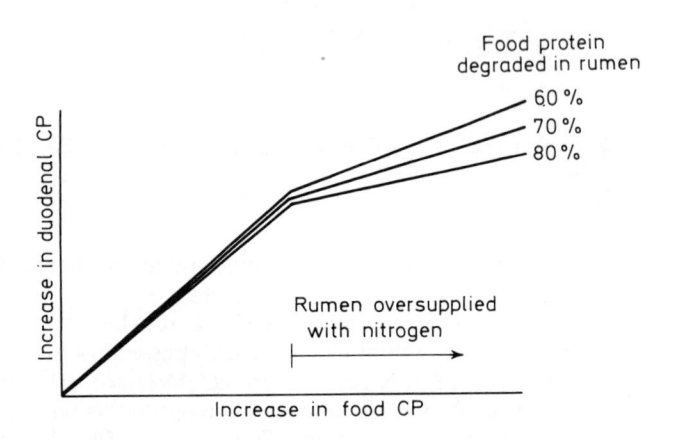

Figure 8.12 The response in duodenal protein supply to increases in food protein supply when the rumen is undersupplied or oversupplied with nitrogen. Response curves are shown for foods of rumen degradability 60, 70 and 80%

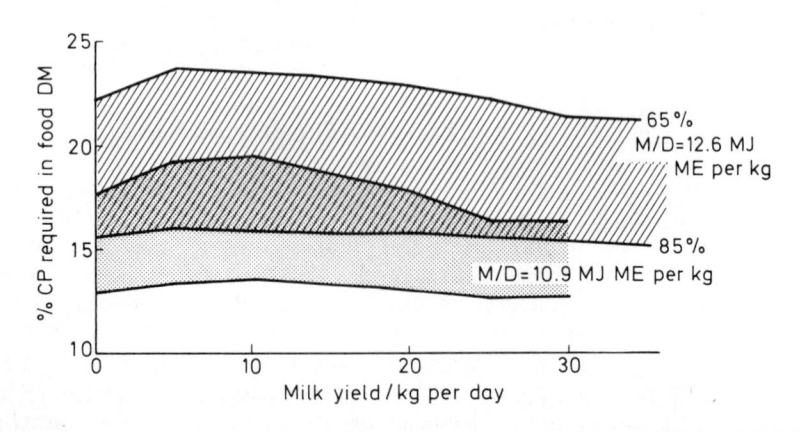

Figure 8.13 Predicted requirements for protein when the efficiency of utilisation of absorbed protein ranges from 65 to 85%. See also Figure 8.11 and text

microbial protein per unit degraded food protein + response to undegraded food protein). Above the limiting value the response is due to undegraded food protein alone. The difference in responses is shown in *Figure 8.12*. When the limiting CP is exceeded the greatest response is achieved with proteins which are relatively resistant to rumen degradation.

To give an impression of the effect which wastage of protein in the rumen can have on predicted CP requirements *Figure 8.13* shows the range of CP % required for a ration of 60% rumen degradability for the two adopted M/D values. The upper and lower bounds of the estimates refer to 65% and 85% efficiencies of utilisation of absorbed protein. In this figure a requirement for EFN has been assumed. The range is broad, but tends to narrow for high yields — this probably reflects changing ME or DM requirements (ARC, 1965).

UREA AS A PROTEIN REPLACER

Figure 8.13 demonstrates quite clearly the importance of selecting ration components to balance the needs of the rumen, and of the host, in relation to energy input. There is, in this context, the need to consider, not only the possible oversupply of degraded nitrogen to the rumen, but also the situation in which less food protein is degraded than the rumen microbes can utilise. In this circumstance the use of urea can be contemplated (*see* p. 143).

Below the CP % in the dry matter which limits microbial protein production urea may be used to maximise microbial protein production. The proportion of CP which can be fed as urea is again dependent on M/D (on account of effects on DM intake), on the efficiency of retention of absorbed protein — at higher efficiencies microbial protein represents a greater proportion of requirement so more urea can be used — and, of course, on food protein degradability. The calculations from which *Figure 8.11* was derived have been used to provide *Figure 8.14* as an example. The inclusion rate of urea N was calculated as the deficit between maximum microbial N production and the supply of degraded food protein N in the rumen. From our calculations there may be scope for urea feeding even for high milk yields but this will depend on the variables described. We cannot make a simple advisory statement on the role of urea but, from *Figure 8.14*, a rule of thumb may be that no more than 20% of food CP should be fed as urea. It could be considerably less under some circumstances.

Protein Intake and Change of Body Reserves

Attention has by and large been riveted on milk production to the exclusion of other aspects of dairy cow performance. The experiments by Gordon and Forbes (1973) (*Figure 8.5*) revealed little effect on live weight gain from variation in protein intake at low energy intakes but an effect at high energy intakes, i.e. the same responses as occurred in milk

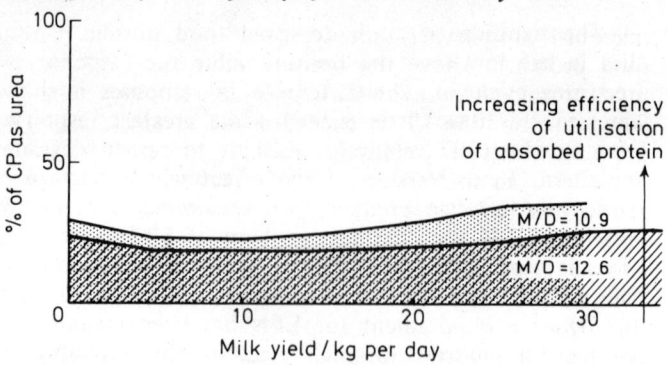

Figure 8.14 The proportion of food CP which can be fed as urea for various milk yields, for foods of different ME concentration (M/D) and for different efficiencies of utilisation of absorbed protein. It was assumed that food protein degradability in the rumen = 60%

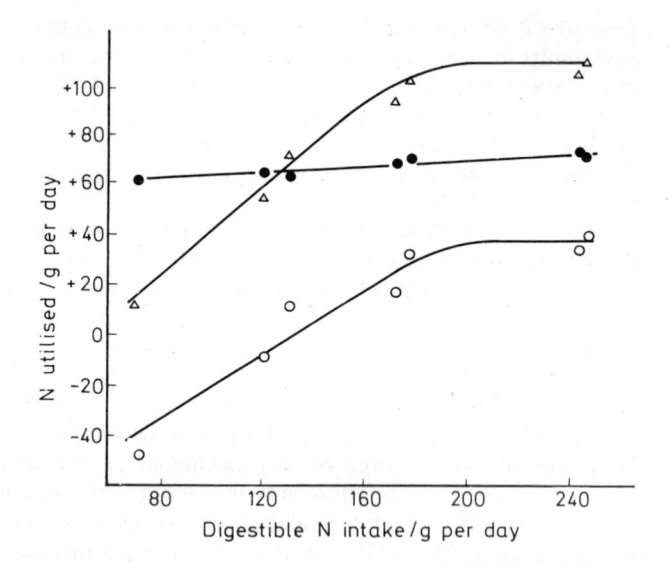

Figure 8.15 The response in dairy cows to digestible nitrogen in the diet (after Balch and Campling, 1961). ○ *body nitrogen exchange;* ● *milk nitrogen excretion;* ▲ *total nitrogen output*

yield. There was a major beneficial effect from higher energy intakes on live weight gain. Balch and Campling (1961) plotted milk nitrogen secretion and body nitrogen retention at various intakes of apparently digestible nitrogen, showing the response to extra nitrogen occurs mainly in body nitrogen retention (*Figure 8.15*). Paquay, de Baere and Lousse (1972) estimated 15 kg protein as the likely lower limit of protein that can be lost from the body, and a recovery period extending over 6 months or more after underfeeding.

The significance of the relationship of protein intake to body changes lies in the long-term nature of milk production. The cow functions in a recurring cycle of gestation—lactation—gestation and so on. Lenkeit (1972) stressed this issue from a long series of experiments on protein utilisation and Broster (1974) stressed it from the point of view of total plane of nutrition. Reid, Moe and Tyrrell (1966) pointed out the possibly erroneous conclusions from short-term trials in which changes of body protein reserves could mask effects of level of protein feeding on milk production. This is especially true in early lactation, a period notoriously difficult to investigate but critical to the success of long-term performance (Broster, 1974).

All this indicates two issues about which there is not adequate information at present: (1) the amount of labile body reserves of the cow; (2) the long-term as opposed to the short-term effects of variation in protein intake.

Huber (1975) made the point that the cow yielding 34 kg milk in early lactation and consuming 21 kg food, needs 15.3% crude protein. This intake provides 21 MJ net energy below NRC (1971) standards. If protein is 20% of tissue energy then the 1 kg mobilised to furnish 21 MJ is equivalent to 0.2 kg protein and a body loss of 30 g N per day, reducing crude protein needed in the diet from 15.3% to 14.4%. For first calf cows Broster, Tuck and Balch (1964) found no advantage from 1.0 kg digestible crude protein per day in late pregnancy compared with 0.7 kg per day so far as subsequent milk yield was concerned, though solids-not-fat content of the milk was improved.

It is implied from the results of Drori and Folman (1970) that cumulative adverse effects do not occur from protein shortage. Such a result was apparent in Broster's (1972) survey of long-term experiments. Thus traumatic effects from small protein shortages need not be looked for.

Future refinements of the metabolic approach to assessment of protein requirements must make allowances for changes in bodyweight during the lactation cycle.

General Discussion

In conclusion a number of points should be emphasised which pertain to:

1. The results of the described approach to defining protein requirements and
2. The future development of the approach.

Our approach has been to describe the requirement as the sum of milk protein output plus an allowance for maintenance (EUN plus truly endogenous faecal N). Some estimates are brought together in *Table 8.8*. At high milk yields the results of our calculations agree quite well with published estimates. Estimates are particularly high for maintenance and low milk yields, and are generally higher than ARC (1965) figures. This could mean that we are underestimating microbial protein production for these ME intakes. These calculations suggest that undegraded food protein

148 *Protein quantity and quality for the UK dairy cow*

Table 8.8 Required CP % in dry matter in relation to milk yield for rations of two M/D values (MJ ME kg^{-1}) and for two efficiencies of utilisation of absorbed protein of 60% degradability in the rumen

	Milk yield (kg per day)			
	0	10	20	30
		CP%		
M/D = 10.88				
0.85 Efficiency	13.0	13.6	13.0	12.6
0.75 Efficiency	14.5	15.1	14.5	14.0
ARC (1965)	5.9	11.4	12.6	12.9
Miller (1973b)	11.3	13.3	14.4	14.5
Broster (1972)	7.1	11.9	15.8	15.7
M/D = 12.55				
0.85 Efficiency	15.5	15.9	15.8	15.3
0.75 Efficiency	17.3	17.7	17.6	17.1
ARC (1965)	6.6	13.0	14.9	15.3

may be required even at maintenance, which is contrary to much of current thinking (Broster and Bines, 1974; Ørskov, 1976). It may also suggest that maintenance should be treated differently from production requirements — if ARC (1965) estimates are correct, then agreement with our approach is reached at an efficiency of utilisation of absorbed protein of > 100%. What this means is not clear. Elliott and Topps (1963) however achieved stability of live weight with diets containing 6.5% CP in DM. Their evidence, and that of others, is incontrovertible.

In all our calculations it has been assumed that microbial protein production can be maximised in relation to defined ME intake by judicial choice of dietary nitrogen components. Choice of feedstuffs is influenced by economic as well as nutritional considerations. It should be simple to apply least cost ration formulation to balance a ration for microbial protein production and undegraded food protein supply.

The importance of energy intake in predicting protein requirements has been stressed and there is also the effect of the ME concentration in the diet. There is a lot of information about protein–energy interactions within the rumen but very little on possible interactions at tissue level. For this we have been forced to take account of any imbalance between absorbed protein and energy supplies within the term 'efficiency of utilisation of absorbed protein'. This is unsatisfactory and future sophistication of the approach will depend on increased knowledge of factors which affect the utilisation of protein within the body. Part of this knowledge must be a description of the maintenance process. Can maintenance be treated in the same way as milk production and how should it be quantitatively described? It is logical that both EUN and endogenous faecal N should be treated as functions of the body rather than of the food. Good estimates of endogenous faecal N are urgently required.

A major strength of considering protein supply as microbial protein plus undegraded food protein is that the place of urea as a feeding supplement can be accurately defined. When degraded food protein supply in the rumen is known urea can be fed to maximise microbial protein production in relation to energy supply. But, at present, degradability characteristics of foods are few and poorly defined. This is another major area where much more information is needed to improve the accuracy of descriptions of requirements. Such information would also lead to better understanding of responses to increased protein inputs — it should be possible to predict a response either in the rumen and in the body, or in the body alone.

A revision of terms is necessary to avoid the inaccuracies of conventional descriptions of protein requirements. Digestible crude protein is meaningless in the present description of protein requirements. It should be abandoned. Protein allowances are described as g CP per day, but as ME and DM intake will always be defined, requirements can be expressed as CP % in DM. Biological value should be replaced by a term which relates to duodenal protein supply or by a value for the digestibility of duodenal protein and an 'efficiency of utilisation of absorbed protein'. Metabolic faecal nitrogen should be replaced by a term for truly endogenous faecal N and related to body size.

The suggested metabolic scheme for predicting protein requirements is robust because it is based on sound descriptive principles. It is resilient because new knowledge can readily be absorbed into its structure without destroying it. Feeding trials, by measuring inputs and outputs, can take the problem of protein requirements only so far, and, notably, variation in results between experiments cannot be analysed in depth. The metabolic approach is an incisive tool for predicting optimal amounts of protein for the dairy cow. The framework is established and greater precision will be achieved as knowledge grows. Adoption of the approach into practice can now be sought.

Acknowledgements

We wish to thank Drs Gordon and Forbes for permission to quote unpublished experimental results; and we should like to thank many colleagues at this Institute and elsewhere, especially Dr. J.H.B. Roy and Dr. R.H. Smith for helpful discussions. We are grateful to the ARC Committee on Nutrient Requirements of Ruminants for permission to participate in some of their discussions on protein utilisation, which have clarified our thinking and influenced the approach made in this paper.

References

Agricultural Research Council (1965). *The Nutrient Requirements of Farm Livestock, No. 2, Ruminants.* London; HMSO

Armstrong, D.G. and Prescott, J.H.D. (1971). In *Lactation*, pp.349–377. Ed. by I.R. Falconer. London; Butterworths
Balch, C.C. (1967). *Wld Rev. Anim. Prod.*, **3**, 84
Balch, C.C. (1972). In *Handbuch der Tierernährung*, Vol. 2, pp.259–291. Ed. by W. Lenkeit, K. Breirem and E. Crasemann. Hamburg; Paul Parey
Balch, C.C. and Campling, R.C. (1961). *J. Dairy Res.*, **28**, 157
Balch, C.C. and Campling, R.C. (1962). *Nutr. Abstr. Rev.*, **32**, 669
Beever, D.E., Coelho da Silva, J.F., Prescott, J.H.D. and Armstrong, D.G. (1972). *Br. J. Nutr.*, **28**, 347
Bines, J.A. and Broster, W.H. (1974). *Proc. Br. Soc. Anim. Prod.*, **3**, 51
Black, J.L. and Griffiths, D.A. (1975). *Br. J. Nutr.*, **33**, 399
Black, A.L., Egan, A.R, Anand, R.S. and Chapman, T.E. (1968). In *Isotope Studies on the Nitrogen Chain*, p.247. Vienna; IAEA
Blaxter, K.L. (1967). *The Energy Metabolism of Ruminants* (2nd edn.). London; Hutchinson
Blaxter, K.L. (1976). Personal communication
Breirem, K. (1949). *Proc. XII Int. Dairy Congr.*, **1**, 28
Broster, W.H. (1972). In *Handbuch der Tierernährung*, Vol. 2, pp.292–322. Ed. by W. Lenkeit, K. Breirem and E. Crasemann. Hamburg; Paul Parey
Broster, W.H. (1973). *Proc. Nutr. Soc.*, **32**, 115
Broster, W.H. (1974). *Bienn. Rev. natn. Inst. Res. Dairy*, 14
Broster, W.H. and Bines, J.A. (1974). *Proc. Br. Soc. Anim. Prod.*, **3**, 59
Broster, W.H., Tuck, V.J. and Balch, C.C. (1964). *J. agric. Sci., Camb.*, **63**, 51
Broster, W.H., Balch, C.C., Bartlett, S. and Campling, R.C. (1960). *J. agric. Sci., Camb.*, **55**, 197
Broster, W.H., Tuck, V.J., Smith, T. and Johnson, V.W. (1969). *J. agric. Sci., Camb.*, **72**, 13
Broster, W.H., Sutton, J.D., Bines, J.A., Corse, D.A., Johnson, V.W., Smith, T. and Jones, P.A. (1971–2). *Rep. natn. Inst. Res. Dairy*, 1971–2, p.74
Broster, W.H., Sutton, J.D., Bines, J.A., Corse, D.A., Johnson, V.W., Smith, T., Siviter, J.W., Napper, D.J. and Broster, V.J. (1973–4). *Rep. natn. Inst. Res. Dairy*, 1973–4, p.67
Burris, W.R., Boling, J.A., Bradley, N.W. and Ludwick, R.L. (1974). *J. Anim. Sci.*, **39**, 818
Burroughs, W., Nelson, D.K. and Mertens, D.R. (1975). *J. Dairy Sci.*, **58**, 611
Buttery, P.J. (1976). In *Principles of Cattle Production*, pp. 145–168. Ed. by H. Swan and W.H. Broster. London; Butterworths
Campling, R.C. (1964). *Proc. Nutr. Soc.*, **23**, 80
Campling, R.C. and Murdoch, J.C. (1966). *J. Dairy Res.*, **33**, 1
Clark, J.H. (1975). *J. Dairy Sci.*, **58**, 1178
Clarke, E.M.W., Ellinger, G.M. and Phillipson, A.T. (1966). *Proc. R. Soc. Series B* **166**, 63
Coelho da Silva, J.F., Seeley, R.C., Thomson, D.J., Beever, D.E. and Armstrong, D.G. (1972a). *Br. J. Nutr.*, **28**, 43

Coelho da Silva, J.F., Seeley, R.C., Beever, D.E., Prescott, J.H.D. and Armstrong, D.G. (1972b). *Br. J. Nutr.*, **28**, 357
Cuthbert, N.H., Thickett, W.S. and Wilson, P.N. (1973). *Proc. Br. Soc. Anim. Prod.*, **2**, 70
Czerkawski, J.W. (1975). *Proc. Nutr. Soc.*, **34**, 62A
Drori, D. and Folman, Y. (1970). *Proc. 18th Int. Dairy Congr.*, p.84
Elliott, R.C. and Topps, J.H. (1963). *Br. J. Nutr.*, **17**, 549
Elliott, R.C., Reed, W.D.C. and Topps, J.H. (1964). *Br. J. Nutr.*, **18**, 519
Evans, E., Buchanan-Smith, J.G., MacLeod, G.K. and Stone, J.B. (1975). *J. Dairy Sci.*, **58**, 672
Frederiksen, L., Østergaard, P.S., Eskedal, H.W. and Steensberg, V. (1931). 136 *Beretn. Forsøgslab.*
Frens, A.M. and Dijkstra, N.D. (1959). *Versl. landbouwk. Onderz.*, Wageningen, 65.9
Gordon, F.J. and Forbes, T.J. (1973). Unpublished
Griffiths, T.W., Spillane, T.A. and Bath, I.H. (1973). *J. agric. Sci., Camb.*, **80**, 75
Hagemeister, H. and Kaufmann, W. (1974). *Kieler Milchw. Forschungsb.*, **26**, 199
Hagemeister, H. and Pfeffer, E. (1973). *Z. Tierphysiol. Tierernährg. Futtermittelk.*, **31**, 275
Harrison, D.G., Beever, D.E., Thomson, D.J. and Osbourn, D.F. (1975). *J. agric. Sci., Camb.*, **85**, 93
Hewitt, C. (1975). *Svensk Veterinärtidning*, **16**, 663
Hogan, J.P. (1973). *Aust. J. agric. Res.*, **24**, 587
Hogan, J.P. and Weston, R.H. (1970). In *Physiology of Digestion and Metabolism in the Ruminant*, pp.474–485. Ed. by A.T. Phillipson. Newcastle upon Tyne; Oriel Press
Hogan, J.P. and Weston, R.H. (1971). *Aust. J. agric. Res.*, **22**, 951
Huber, J.T. (1975). *J. Anim. Sci.*, **41**, 954
Hume, I.D. (1970a). *Aust. J. agric. Res.*, **21**, 297
Hume, I.D. (1970b). *Aust. J. agric. Res.*, **21**, 305
Hume, I.D. and Purser, D.B. (1975). *Aust. J. agric. Res.*, **26**, 199
Jackson, P. (1974). *Proc. 8th Nutr. Conf. Feed Mfrs.*, University of Nottingham, pp.123–142. Ed. by H. Swan and D. Lewis. London; Butterworths
Jakobsen, P.E. (1957). 299 *Beretn. Forsøgslab.*
Kaufmann, W. and Hagemeister, H. (1975). *Übers. Tierernährg.*, **3**, 33
Klooster, A. Th. van't and Rogers, P.A.M. (1969). *Meded. LandbHoogesch. Wageningen.*, **69-11**, 3
Leibholz, J. (1972). *Aust. J. agric. Res.*, **23**, 1073
Lenkeit, W. (1972). In *Festskrift Til Knut Breirem*, pp.123–140. Ed. by L.S. Spildo, T. Homb and H. Hvidsten. Gjovik; Mariendals Boktrykkeri As
Lindsay, J.R. and Hogan, J.P. (1972). *Aust. J. agric. Res.*, **23**, 321
MacRae, J.C. and Ulyatt, M.J. (1974). *J. agric. Sci., Camb.*, **82**, 309
MacRae, J.C., Ulyatt, M.J., Pearce, P.D. and Hendtlass, J. (1972). *Br. J. Nutr.*, **27**, 39
McAllan, A.B. and Smith, R.H. (1972). *Proc. Nutr. Soc.*, **31**, 24A

McAllan, A.B. and Smith, R.H. (1974). *Br. J. Nutr.*, **31**, 77
McGilliard, A.D. (1961). 'Re-entrant duodenal fistula techniques: application to the study of digestion and passage in the bovine alimentary tract.' Ph.D. thesis. Michigan State University
Mangan, J.L. (1972). *Br. J. Nutr.*, **27**, 261
Mason, V.C. (1969). *J. agric. Sci., Camb.*, **73**, 99
Mathison, G.W. and Milligan, L.P. (1971). *Br. J. Nutr.*, **25**, 351
Miller, E.L. (1973a). *Proc. Nutr. Soc.*, **32**, 79
Miller, E.L. (1973b). Quoted by Bines and Broster (1974)
Moe, P.W., Reid, J.T. and Tyrrell, H.F. (1965). *J. Dairy Sci.*, **48**, 1053
Murdoch, J.C. (1962). *J. Br. Grassld Soc.*, **17**, 268
National Research Council (1971). *Recommended Dietary Allowances.* (7th edn.). Washington, DC; Natl. Acad. Sci.
Nicholson, J.W.G. and Sutton, J.D. (1969). *Br. J. Nutr.*, **23**, 585
Nolan, J.V. and Leng, R.A. (1972). *Br. J. Nutr.*, **27**, 177
Oldham, J.D. (1973). 'Dietary carbohydrate and nitrogen interactions in the rumen.' PhD Thesis, University of Nottingham
Ørskov, E.R. (1976). In *Protein Metabolism and Nutrition*, pp.457–476. Ed. by D.J.A. Cole, K.N. Boorman, P.J. Buttery, D. Lewis, R.J. Neale and H. Swan. London; Butterworths
Ørskov, E.R. and Fraser, C. (1973). *Proc. Nutr. Soc.*, **32**, 68A
Ørskov, E.R., Fraser, C. and McDonald, I. (1971). *Br. J. Nutr.*, **25**, 243
Ørskov, E.R., Fraser, C. and McDonald, I. (1972). *Br. J. Nutr.*, **27**, 491
Ørskov, E.R., Fraser, C., McDonald, I. and Smart, R.I. (1974). *Br. J. Nutr.*, **31**, 89
Ørskov, E.R., McDonald, I., Fraser, C. and Corse, E.L. (1971). *J. agric. Sci., Camb.*, **77**, 351
Paquay, R., de Baere, R. and Lousse, A. (1972). *Br. J. Nutr.*, **27**, 27
Paquay, R., Godeau, J.M., de Baere, R. and Lousse, A. (1973). *J. Dairy Res.*, **40**, 329
Pfeffer, E., Kaufmann, W. and Dirksen, G. (1972). *Z. Tierphysiol. Tierarnähr. Futtermittelk.*, Suppl 1, 22
Pilgrim, A.F., Gray, F.V., Weller, R.A. and Belling, C.B. (1970). *Br. J. Nutr.*, **24**, 589
Purser, D.B. (1970). *Fedn. Proc. Fedn. Am. Socs. exp. Biol.*, **29**, 51
Purser, D.B. and Buechler, S.M. (1966). *J. Dairy Sci.*, **49**, 81
Reid, J.T., Moe, P.W. and Tyrrell, H.F. (1966). *J. Dairy Sci.*, **49**, 215
Robinson, J.J. and Forbes, F.J. (1970). *Anim. Prod.*, **12**, 601
Rook, J.A.F. (1961). *Dairy Sci. Abstr.*, **23**, 251 and 303
Rook, J.A.F. and Line, C. (1961). *Br. J. Nutr.*, **15**, 109
Rook, J.A.F. and Line, C. (1962). *Proc. XVI Int. Dairy Congr.*, 1:1, 57
Rowland, S.J. (1946). *Dairy Ind.*, **11**, 656
Roy, J.H.B. (1975). Personal communication
Satter, L.D. and Roffler, R.E. (1975). *J. Dairy Sci.*, **58**, 1219
Satter, L.D. and Slyter, L.L. (1974). *Br. J. Nutr.*, **32**, 199
Sharma, H.R., Ingalls, J.R. and Parker, R.J. (1974). *Can. J. Anim. Sci.*, **54**, 305
Smith, R.H. and McAllan, A.B. (1974). *Br. J. Nutr.*, **31**, 27
Smith, T. (1976). Unpublished

Stobo, I.J.F. and Roy, J.H.B. (1973). *Br. J. Nutr.*, **30**, 113
Stouthamer, A.H. and Bettenhausen, C. (1973). *Biochimica et Biophysica Acta*, **301**, 53
Sutton, J.D. (1976). In *Principles of Cattle Production*, pp.121–143. Ed. by H. Swan and W.H. Broster. London; Butterworths
Sutton, J.D., Smith, R.H., McAllan, A.B., Storry, J.E. and Corse, D.A. (1975). *J. agric. Sci., Camb.*, **84**, 317
Tagari, H., Krol, O. and Bondi, A. (1965). *Nature*, **206**, 37
Tagari, H., Ben Gedalya, D., Shevach, Y. and Bondi, A. (1971). *J. agric. Sci., Camb.*, **77**, 413
Tamminga, S. (1975). *Neth. J. agric. Sci.*, **23**, 89
Thomas, J.W. (1971). *J. Dairy Sci.*, **54**, 1629
Van Es, A.J.H. (1972). In *Handbuch der Tierernährung*, Vol. 2, pp.2–54 Ed. by W. Lenkeit, K. Breirem and E. Crasemann. Hamburg; Paul Parey
Wagner, D.C. and Loosli, J.K. (1967). *Mem. Cornell Agric. Exp. Stn.* No. 400
Watson, M.J., Savage, G.P. and Armstrong, D.G. (1972). *Proc. Nutr. Soc.*, **31**, 98A
Weller, R.A. (1957). *Aust. J. biol. Sci.*, **10**, 384
Weller, R.A. and Pilgrim, A.F. (1974). *Br. J. Nutr.*, **32**, 341
Wiktorsson, H. (1971). *J. Dairy Sci.*, **54**, 374
Woodman, H.E. (1957). *Bull. Minist. Agric. Fish. Fd* (14th edn.). No. 48. London; HMSO

9

NUTRITIONAL SYNDROMES OF POULTRY IN RELATION TO WHEAT-BASED DIETS

C.G. PAYNE
Poultry Husbandry Research Foundation, University of Sydney, Australia

Recently nutritional research with poultry in Australia has gained impetus due to government-controlled research levies obtained from both the poultry meat and egg industries. Additionally, the growing realisation that information on nutritional allowances and deficiency syndromes obtained from poultry fed corn—soya type diets often is not relevant when the predominant cereal is wheat and the main protein sources are animal proteins together with less well researched vegetable proteins such as safflower, sunflower, lupinseed and rapeseed.

The Poultry Husbandry Research Foundation is unique in that nutritional cum management problems involving poultry (and occasionally pigs and fish) are referred directly to it by the industry whose Foundation members include feed companies, integrated organisations, breeders, hatcherymen and the larger poultry farmers. In addition various services are carried out such as total amino acid analyses, available lysine determinations, metabolisable energy evaluation and trace mineral analyses (selenium and molybdenum). During 1975, the unit was involved in a series of nutritional problems involving large integrated broiler organisations, which turned out unexpectedly to show molybdenum, pyridoxine and folic acid responsive conditions. The background to these syndromes, which are prevalent in many countries, as well as the peculiarities of some more obscure feed ingredients used in Australia, are described in this chapter. The nutritional syndromes referred to are most common on (but not exclusive to) wheat-based diets, especially when supplemented with meat meal and some of the less well documented vegetable proteins.

Ingredient Composition

Amino acid composition, together with determined metabolisable energy values for some Australian feed ingredients are given in *Table 9.1*. More complete Australian information for a wide range of materials has been published by the Queensland Department of Primary Industries (1974). *Table 9.1* includes fishmeal — 65% protein — samples from South America as a reference for comparison with other laboratories. In our determinations we have refluxed 1g samples in 400 ml 6N HCl under nitrogen at 136°C

Table 9.1 Amino acid composition (g per 16 g N) and metabolisable energy values (KJ per kg) for Australian ingredients. Data in brackets indicate typical protein levels.

Amino acids	Coconut (22%)	Cottonseed (40%)	Fishmeals (65%)	Fishmeal (50%)	Lupins Uniwhite (29%)	Meat and bone meal (44%)	Meat meals (49%)
Aspartic	6.9	7.4	7.7	7.0	7.3	5.7	6.0
Threonine	3.5	3.0	3.0	4.1	2.9	2.9	3.1
Serine	3.9	3.7	3.0	3.4	3.7	3.1	3.6
Glutamic	20.3	20.6	15.0	12.6	21.2	11.8	13.0
Proline	2.9	3.5	3.0	5.4	3.4	7.4*	7.0**
Glycine	4.2	4.4	7.0	10.5	4.2	15.7	13.5
Alanine	4.3	3.7	5.0	6.4	2.7	6.7	6.5
Valine	5.0	4.1	4.7	5.1	3.5	4.2	4.2
Cystine	0.82	1.44	0.85	0.59	0.98	0.54	0.80
Methionine	1.71	1.56	3.18	3.01	0.66	1.59	1.59
Isoleucine	2.7	3.1	3.9	4.4	4.4	2.8	3.2
Leucine	5.7	5.9	8.1	7.1	6.7	6.5	6.5
Tyrosine	2.2	2.9	3.9	3.7	3.7	2.4	2.7
Phenylalanine	4.1	5.2	4.7	4.2	4.2	3.6	3.8
Ammonia	1.2	2.0	1.2	1.2	2.2	1.1	1.0
Lysine	3.0	4.3	8.8	6.0	4.8	5.0	5.6
Histidine	1.9	2.8	3.2	3.4	2.5	2.0	1.9
Arginine	12.9	10.9	6.6	5.9	10.1	7.3	7.1
Tryptophan	0.95	1.25	0.92	0.80	1.21	0.63	0.70
Metabolisable energy	7100	8100	11 800	10 100	9700	9300	10 500

* plus 9.1 – hydroxyproline
** plus 8.4 – hydroxyproline

Table 9.1 (Continued)

Amino acids	Rapeseed (36%)	Rice bran (13%)	Safflower (29%)	Sorghum (9%)	Soyabean (45%)	Sunflower (38%)	Wheat (12%)	High protein wheat (17%)
Aspartic	5.4	7.8	8.0	5.3	9.2	7.5	4.8	3.8
Threonine	4.4	3.4	3.0	3.1	3.8	3.6	2.8	2.4
Serine	3.3	3.5	3.3	3.8	4.0	3.7	3.8	3.5
Glutamic	19.5	13.6	21.3	21.5	18.8	20.4	31.6	32.8
Proline	6.0	3.5	2.3	8.4	4.6	3.9	10.0	9.1
Glycine	5.9	5.5	6.5	4.0	4.7	6.2	4.6	4.8
Alanine	4.3	6.0	3.8	7.9	3.6	4.2	3.3	3.0
Valine	5.0	4.9	4.7	4.9	4.7	4.7	4.5	4.6
Cystine	1.23	1.26	1.27	1.24	1.11	1.14	1.41	1.50
Methionine	2.28	1.82	1.51	1.62	1.46	2.44	1.67	1.67
Isoleucine	4.1	4.2	3.7	4.0	4.9	4.1	3.8	3.7
Leucine	7.6	7.0	6.4	13.0	8.1	6.8	7.9	7.2
Tyrosine	3.2	3.0	3.2	4.5	3.9	2.5	3.3	2.6
Phenylalanine	4.8	4.2	4.8	5.5	5.1	5.1	5.0	4.8
Ammonia	2.1	2.1	2.2	2.4	1.8	2.1	3.4	3.9
Lysine	6.2	4.8	3.3	2.4	6.1	3.4	3.3	2.5
Histidine	2.9	3.0	2.7	2.6	2.6	2.5	2.4	2.5
Arginine	6.2	8.2	9.2	3.6	7.1	7.9	4.9	4.5
Tryptophan	1.83	0.70	1.50	1.30	1.53	1.63	1.40	1.45
Metabolisable energy	7100	12 500	5500	14 000	10 400	8650	13 200	12 600

for 36 hours. Shorter times of hydrolysis do not result in complete amino acid release; smaller samples have inherent sampling errors; and smaller volumes of acid, especially in sealed tubes, often involve destruction of amino acids including threonine.

Cottonseed Meal

Australian data (Packham and Payne, 1973a; Packham, Royal and Payne, 1973b) have shown that lysine availability in cottonseed meals is affected markedly by processing conditions. At high levels of inclusion (31.3%) in broiler diets, cottonseed meals did not depress broiler performance, provided that 0.05% ferrous sulphate was added to the diets to counteract any gossypol toxicity, and the dietary leucine level was kept above 1.25% of the diet. When wheat is the sole cereal in the cottonseed diets, it is difficult to maintain leucine and isoleucine levels, especially when meat meal is combined with cottonseed meal. Under these circumstances isoleucine tends to become limiting to broiler performance particularly if the meat meal contains appreciable quantities of blood.

Lupinseed

The 200 or so wild species of lupins (Flight, 1956) represent a large and almost untapped pool of genetic material for protein ingredients. The use of lupins for stockfeed can be associated with two distinct nutritional problems — lupinosis and lupinine poisoning. Lupinosis is a haemorrhagic disease involving copper metabolism and affects animals grazing lupins. It is thought to involve fungal toxins formed on the lupin plant (Gardiner, 1964; Flight, 1956). Lupin poisoning (lupin madness) is caused by alkaloids (lupinines) in the seeds of wild, 'bitter' strains. Fieser and Fieser (1957) have postulated that the lupinine molecule is formed by two molecules of lysine condensing together; hence 'sweet' varieties tend to have higher lysine contents than do the wild, 'bitter' strains. Gladstones (1971) in Western Australia has described the production of varieties devoid of lupinine. At present the most satisfactory from an agronomic view are 'sweet' varieties of *lupinus angustifolius* (Uniwhite, Unicrop, Uniharvest) which have been selected for uniform early ripening and lack of seed shedding at harvest. The amino acid composition of these varieties is constant, and the extremely low methionine levels might be useful in formulating experimental diets for biological assays for the sulphur-containing amino acids. The seed 'as-is' protein level of 29% may be raised to 36% through removal of the seed coat. Ground seeds and ground dehulled seeds have been included in correctly balanced broiler, layer and pig diets at levels of up to 25% without depression of performance. Miller (1972) fed up to 25% lupinseed to second year hens. Performance was extremely uniform (no significant differences) with the highest rate of lay from the 25% inclusion, and the heaviest eggs from the 16.7% lupinseed diet. Smetana and Morris (1972), and Smetana (1974) have fed up to 22% lupinseed in correctly balanced broiler diets. Although no significant differences in performance existed, again the highest levels of lupinseed were associated with the best performance.

Meat and Bone Meals

These generally are the cheapest Australian protein ingredients; however, especially in broiler diets, inclusion levels have to be limited so that total calcium levels do not exceed 1.3%. At higher dietary calcium levels, performance is depressed partly due to increased severity of some diseases — especially infectious bronchitis (Cumming, 1975) — partly to low and varied available lysine levels, and also to effects of meat and bone meal on trace nutrient requirements — especially pyridoxine and folic acid. Sathe (1964) fed wheat—meat diets containing 44.9% (*sic*) meat meal in an attempt to accentuate differences between meat meals. Presumably the 4.1% total dietary calcium had adverse effects on intestinal B vitamin synthesis, but it was shown that diets based on three different low quality meat meals were deficient in:

1. pyridoxine and folic acid equally.
2. folic acid singly.
3. pyridoxine first, pantothenic acid second, and folic acid third.

The diets may have been highly impractical; however the symptoms, especially those of a joint pyridoxine—folic acid deficiency, are identical to several syndromes which are occurring in very young broiler flocks, and which in Australia are responsive to and preventable by high levels of pyridoxine and folic acid supplementation, particularly in the breeder diets. The symptoms as described by Sathe (1964) are: 'Hypophagia and weakness, usually on the 3rd or 4th days. In some cases wings were held up at the elbow, and the primary feathers drooped markedly. Intermittent nervous tremours in wings were also noticed. Depending on the degree of severity, chicks showed varying symptoms such as sitting on the hocks with or without a rocking motion, head slightly bent downwards, and lying on the abdomen with inability to lift up the neck. Very severely affected chicks were prostrate.'

In Australia, I have induced inadvertently this same pyridoxine—folic acid responsive syndrome in broiler chicks aged 2—3 weeks fed wheat-based diets which also contained 10% meat meal, 10% soyabean meal, 10% sunflower meal, plus a comprehensive supplement including 4 ppm pyridoxine, 2 ppm folic acid and 100 ppm copper as copper sulphate. Doubling the copper sulphate level to 200 ppm Cu resulted in a 100% incidence of this syndrome. The syndrome can be prevented by trebling pyridoxine and folic acid levels. It is interesting to speculate on earlier research which studied the use of copper sulphate as a growth promotant (*see* Fisher *et al.*, 1973) and postulate why pyridoxine and folic acid deficiencies did not occur. However, a gizzard erosion was described by Fisher *et al.* (1973) which became a serious problem when feeding copper levels of 405 and 605 ppm. It was present at a much lower incidence at a dietary copper level of 205 ppm. In a detailed description of pyridoxine deficiency in broiler chickens (Daghir and Shah, 1963), a gizzard erosion was a prominant feature, but was not described in detail.

Rapeseed

Our research unit has compared imported solvent-extracted Canadian 'Oro' with local screw-pressed and solvent-extracted meals. With both broilers and layers, feeding of rapeseed meal has resulted in considerable mortality which has been directly proportional to the level of inclusion and duration of feeding. Examinations *post mortem* (Koentjoko, 1974) have shown lesions as described by Hall (1974) with hepatosis, lysis of the reticulin and massive liver haemorrhages even in broiler chickens aged 5 days. Koentjoko (1974) in a 224 day laying trial obtained 5% mortality from the liver syndrome through feeding 4% of rapeseed meals in layer diets. At 20% inclusion, 46 mortalities of a flock of 283 hens (16.9%) had severe liver lesions — signs of the rapeseed liver syndrome. Because of suggestions that the problems might be due to reactivation of the isothiocyanates and oxazolidinethione by intestinal bacteria, the use of various antibiotics in conjunction with rapeseed was studied in broilers and layers. Incomplete data to date indicate that some antibiotics prevent the increase in thyroid size associated with rapeseed meals, and other antibiotics reduce the incidence of the liver haemorrhages. Overall the data suggest (Van Leong, 1975) that two distinct problems exist. In experiments with layers, Van Leong has incorporated erucic acid into soyabean diets, with no detrimental effects on health or production. These observations together with Canadian work published recently (Olumu *et al.*, 1975) would suggest that the importance of erucic acid may have been overemphasised, and the importance of other toxic factors under-rated.

In considering the usefulness of rapeseed meal for poultry, attention should be given to criteria other than live weight gain and feed conversion efficiency. For example Peters *et al.* (1974) noted a gradual (nonsignificant) increase in mortality from 5.0 to 8.9% of broilers started when rapeseed meal was increased from nil up to 6% in broiler diets. In addition there was almost a linear increase in the incidence of deformed birds as the rapeseed levels were increased from nil up to 9%.

One further major peculiarity in trace nutrient content of rapeseed meals, is a 3% tannin content. Scott, Nesheim and Young (1969) considered that this would limit the amount of rapeseed which could be used in poultry diets to 15%, as chickens appear not to be able to assimilate more than 0.5%. Koentjoko (1974) fed diets containing 20% rapeseed and 0.7% tannin to layers. No effects were observed on production through raising the tannin level to 1.2%; and it was concluded that tannin content represents a minor problem of rapeseed.

Rice Bran

Recent work by Kratzer et al. (1974) with broilers demonstrated that rice bran might contain heat-labile trypsin inhibitors. An Indonesian postgraduate (Major, 1975) has experimented with 0, 30 and 60% inclusion of rice bran in layer diets. At the 60% inclusion, a significant depression in egg production occurred, which was prevented entirely by autoclaving the rice bran for 15 minutes at 15 psi.

Kratzer, Earl and Chiarquanont (1974) reported that there appeared to be no evidence that the growth depression caused by rice bran is related to amino acid or protein availability. Autoclaving and steam treating the rice bran consistently improved its feeding value for chickens. The mechanism for this improvement is unknown, although the pancreatic hypertrophy obtained on rice bran diets appeared similar in the work of Kratzer, Earl and Chiarquanont (1974) to the trypsin inhibitor in soyabeans. Rice bran does not appear to be unique amongst the cereals in containing anti-trypsin factors. Creek et al. (1962) reported an anti-trypsin factor in raw wheat germ, which could be inactivated by autoclaving.

Wheat as the Predominant Cereal

In many reports, the feeding value of wheat is, as expected, that of a fairly high energy cereal. Its metabolisable energy value is consistently reported to be only 6—8% lower than corn (Scott, Nesheim and Young, 1969), with a normal value of 3200—3300 kcal per kg (13 808—13 827 MJ per kg) 'as is'. However in two series of experiments in Australia (Connor and Major, 1975) the metabolisable energy value obtained for wheat (in contrast to sorghum and to wheat bran determination) was consistently lower by about 10% (range 7—25%) when the level of inclusion in test diets was higher than 70%. At 50% or lower levels of inclusion, normal metabolisable energy values were obtained. These data are contrary to the accepted theory of metabolisable energy additive values, and support the idea that wheat may contain a slightly toxic inhibitor.

On the other hand high protein wheats have been fed to laying hens to provide all the energy and protein, at levels in excess of 85%. Provided methionine and lysine levels were adjusted correctly, egg production was maximised (Turner and Payne, 1971).

In terms of trace nutrient levels, wheat shows many differences compared to corn and sorghum. Some of the problems which have resulted when wheat has been included in lieu of corn in poultry diets have almost certainly been a result of these differences. The main difference in trace mineral content between wheat and corn is probably in manganese, with wheat having approximately twice the content of corn (Scott, Nesheim and Young, 1969; McDonald et al., 1969).

In vitamin content wheat normally contains appreciably less biotin, pyridoxine and vitamin E, but more nicotinic acid, pantothenic acid and choline than does corn (Scott, Nesheim and Young, 1969). Scott (1966) suggested that the nicotinic acid and pantothenic acid in wheat may be largely unavailable to poultry. Wheat also contains considerably less linoleic acid. These differences in trace nutrient content help to explain several syndromes including the fatty liver and kidney syndrome (FLKS) and its high incidence on wheat diets; the pyridoxine-responsive hatchery syndrome (described later in this paper); and the depressions in egg size attributed to inadequate linoleic acid in wheat layer diets. Biotin deficiencies are also common in turkeys especially when they are fed wheat—animal protein diets. Robblee and Clandinin (1970), who described this syndrome, showed that the total requirements for turkey poults were 0.22 ppm biotin, 3.3 ppm folic acid and 11 ppm pantothenic acid. These requirements are higher than had the turkeys been reared on corn—soya diets.

A simple comparison of the basal levels of trace nutrients in two diets, 80% corn, 20% soyabean and 80% wheat, 20% meat meal is given in *Table 9.2*. The main inadequacies of the wheat—meat diet compared to the corn—soya diet are pyridoxine, folic acid, biotin and vitamin E. However in terms of basal mineral levels the wheat—meat diet is richer in iron and manganese and lower in potassium. Such differences in basal trace nutrient levels should result in compensatory variations in trace nutrient supplements.

Table 9.2 Estimated trace nutrient levels (ppm) resulting from the basal ingredients in corn—soya and wheat—meat diets. Trace nutrient levels are based on McDonald *et al.* (1969), Scott, Nesheim and Young (1969) and personal observation

Nutrient	Requirement (NRC 1971) Broilers	Breeders	Estimated basal levels 80% corn 20% soya	80% wheat 20% meat
Copper	4	?	13	9
Iron	80	?	60	260
Magnesium	500	?	1800	1800
Manganese	55	33	13	40
Potassium	2000	?	6800	4400
Zinc	50	65	36	39
Biotin	0.090	0.150	0.125	0.070
Choline	1300	?	1050	1200
Folic acid	0.55	0.35	0.64	0.22
Nicotinic acid	27	10	23	60
Pantothenic acid	10	10	7.5	10
Pyridoxine	3.0	4.5	7.2	2.0
Vitamin B_2	3.6	3.8	1.5	2.0
Vitamin E	10	?	19	9

Apparent Changes in Nutrient Requirements

The demonstration that FLKS in broilers is biotin-responsive (Payne et al., 1974) exposed the fallacy of believing that no major developments in trace nutrient research could be expected. It is likely that some trace nutrient levels in crops will change, especially as a result of monocropping, but also due to new varieties, changes in agronomic techniques (such as fertiliser levels and irrigation), variations in soil pH, and many other factors. For example it is known that the molybdenum uptake by plants decreases dramatically as soil pH falls (Black et al., 1965); simultaneously manganese uptake increases markedly. Hence small changes in soil pH can result in wide variation in crop mineral ratios. Various other factors (see Black et al., 1965) affect molybdenum uptake by plants: elevated sulphate levels reduce molybdenum uptake, and increased phosphate levels have the opposite effect even in the absence of soil pH changes.

With genetic selection for improved performance in meat animals, especially associated with more rapid deposition of proteins in body tissues, it is likely that the requirements of meat animals for several trace nutrients will change. Enzymes especially involved in deposition of body proteins include pyridoxine, nicotinic acid and folic acid. It is possible that as genetic improvements in growth rate increase, there may be disproportionate elevations in the requirements of these vitamins. Wide genetic differences in requirements for poultry have been reported for several trace nutrients such as folic acid (Lillie and Brigg, 1947), pyridoxine (Lucas, Heuser and Norris, 1946) and riboflavin (Culton and Bird, 1940). Although there are likely to be changes in requirements of meat animals for some trace nutrients, especially as improvements in performance occur, the requirement changes of the parental stock may not differ in the same nutrients, or to the same amount. Broiler breeding stock will continue to consume the same quantity of food per day. However because growth rate improvements are due basically to ability to consume voluminous quantities of food, the food intake capacity of meat parental stock will increase. Thus their actual intake expressed as a proportion of potential intake under *ad libitum* dietary regimes will diminish, and they will consume their daily packets of food more quickly. In addition to management problems, it is likely that the periodic feeding, with daily periods of relative starvation, may alter internal intestinal conditions both of pH and flora. Thus some of the reproductive disorders, both in sows and broiler breeding stock (which have been studied at Sydney University recently) appear to be due to a reduction in synthesis of vitamins such as biotin, pyridoxine, K, and folic acid, where parts of the animal's 'requirements' are provided by dietary inclusion and parts are provided by intestinal synthesis. These vitamin-responsive reproductive disorders have most commonly occurred when the breeder diets have contained abnormally high levels of trace minerals (especially as sulphates). Presumably there may have been antibiotic-like effects paralleling that of the growth promotant, copper sulphate. In addition, where integrated broiler breeding organisations have had broiler breeding stock housed in several locations, poor hatchability problems have been worst where the drinking water has been hard, particularly when the sulphate levels have been higher than 50 ppm. Recently poor hatchability problems have occurred in many countries, both in turkeys (Wallis and Parker, 1974) and in broiler breeding flocks (Cruickshank, 1975). In the turkeys, symptoms included 'clubbed' down amongst dead in shell embryos associated with subnormal hatchability. Wallis and Parker (1974) suggested that the condition might have been due to the use of poor quality protein ingredients of relatively unknown trace nutrient composition. In the field report by Cruickshank (1975) it was suggested that vitamins K, B_2, B_{12}, pantothenic acid, choline, zinc, and manganese were perhaps deficient. No mention was made of the nutrients which in the Sydney University research were shown to be inadequate — molybdenum, pyridoxine and folic acid. Cruickshank (1975) reported that 'the problems persisted even when the nutrient levels have been increased'. It should be noted that *ad hoc* supplementation with some trace nutrients (but not including the deficient one(s)) will at the best cause no change in any problem, but because of trace nutrient imbalances the situation generally will be worsened by such a casual approach.

Increased Requirements for Trace Nutrients?

As discussed later, the necessity for dietary pyridoxine and folic acid supplementation appears to be increasing. Factors which may be involved in these increased allowances include:

1. Increasing requirements?
2. Reduced intestinal synthesis?
3. Increased levels of antagonists, inhibitors, etc?
4. Underestimation of requirement when interpreting research.
5. Nutrients being in unavailable forms.
6. Nutrients being made unavailable by other trace nutrients.

Many of the vitamin allowances for breeding stock were ascertained over short-term experiments, e.g. Fuller (1964) worked on the pyridoxine requirements of breeding stock over 8 weeks. Nutrient standards often appear to have disregarded information which suggests higher nutrient requirements than the majority of work suggests. Thus for broiler chickens most standards suggest a requirement for pyridoxine below 4 ppm. However higher estimates have been provided (yet subsequently largely ignored) by Lucas, Heuser and Norris (1946) (well in excess of 5 ppm); Kirchgessner and Friesecke (1963) ($> 5 < 10$ ppm); Daghir and Shah (1973) (> 5.1 ppm); and Yoshida and Morimoto (1960) (*c.* 6 ppm). Pyridoxine requirements and allowances are further complicated in that pyridoxine assays using bacteria give equal value to pyridoxine, pyridoxal and pyridoxamine. However there is information (Sarma, Snell and Eluehjem, 1946) that for poultry the availability of pyridoxal and pyridoxamine is low, due to competition from intestinal flora especially *lactobacilli*.

Excessive intakes of iron salts can cause nutritional disturbances (Scott, Nesheim and Young, 1969) due to formation of colloidal insoluble iron phosphates, which may absorb vitamins and trace minerals thereby preventing their adsorption. Additionally an insoluble complex between ferric chloride and pyridoxine may be formed. It is possible that trace nutrient excesses could create inositol deficiencies particularly in poultry breeding stock. Inositol is normally produced from glucose or pyruvate pathways; however in incubating eggs these do not seem to exist (see Alam, 1971). Inositol can also be synthesised from phytic acid which is the hexa phosphoric ester of inositol. Phytic acid forms insoluble complexes with many trace minerals (thereby increasing their requirements); hence deficiencies of inositol might occur due to excessive trace mineral supplementation.

Trace Mineral Interactions

Interactions of trace minerals are more prominent than amino acid interactions, and tend to be more complex. The more important interactions have been detailed by Buck and Evans (1973), Roine and Ahlström (1970) and Davies (1974). Because of widespread molybdenum deficiencies in poultry, mineral interactions which affect the incidence and severity of these molybdenum-responsive conditions are summarised in *Table 9.3*.

Table 9.3 Interactions involving molybdenum in non-ruminant nutrition

Nutrients which may increase the apparent molybdenum requirement	
Tungstates	Not as effective antagonists as copper sulphate.
Copper sulphate } Arsenicals	When used at growth promotant levels, these often create accidentally molybdenum-responsive conditions
Total dietary transitional element content	Presumably by competition both for absorption sites, and at the cellular level.
Total sulphate content	Especially significant when permanent hardness in drinking water (*see* text).
Nutrients which may alleviate molybdenum inadequacies	
Selenium } Vanadium	(*See* text).
Nutrients which may be made 'more available' with molybdenum supplementation	
Fluorine	See papers summarised by Davies (1971).
Zinc	Mo supplementation may reduce the quantity of zinc needed to prevent parakeratosis (*see* text).
Iron	See paper by Seelig (1972).

Selenium deficiency may be an occasional problem specific to ruminants in a few paddocks in Australia. Selenium levels in Australian feed ingredients (Nell, 1975) do not suggest that selenium deficiency could occur in pigs or poultry fed multi-ingredient diets. However a reproductive syndrome in sows (Hartley, 1975) and a syndrome in broiler breeding stock described by Bains, Mackenzie and McKenzie (1975) were each shown to be selenium-responsive. This latter condition (Bains, 1975) was reduced markedly but not eliminated by selenium medication via the drinking water. In the affected flocks, adult birds were still being affected by the condition at the same time as selenium toxicity symptoms were appearing in hatchery dead in shell. This observation would suggest that the selenium responsive condition was not a prima facie selenium deficiency. In addition the condition was only reported on one of two large sites. The only difference noted between the feeding, management and stock on these two sites was in the drinking water. Affected flocks consumed hard water which contained about 60 ppm sulphate within 333 ppm total solids. Unaffected flocks had water with no sulphate and only 100 ppm total solids, 96% of which was attributable to carbonates. Once molybdenum additions were made to the broiler breeder diets, then the selenium-responsive condition disappeared. It is at the cell level that mineral interactions are most important. A table has been published of the involvement of metal ions with approximately 220 enzymes (Davies, 1971). From the data it may be calculated that copper is found as an essential component of 12 of these enzymes and copper integrates and/or activates a further 5 enzymes. However an excess of copper inhibits 12 enzymes. For zinc the figures are: essential component, 8; integrator—activator, 13; inhibitor, 12. These observations stress the necessity of having the correct balance of trace minerals both in the diet and at the cellular level.

Molybdenum Deficiencies in Poultry and Pigs

My research at Sydney University has shown that two conditions, a 'pseudo clubbed down, ginger hair hatchery syndrome' and the 'scabby hip and femoral bone syndrome', are prevented by dietary molybdenum supplementation. In addition there is circumstantial evidence to suggest that molybdenum supplementation to pig diets may be beneficial in increasing reproductive efficiency of sows, and in alleviating parakeratosis. Where copper sulphate at the growth promotant level (100–200 ppm Cu) has been accidentally fed to gilts prior to first mating, anoestrous, small litter size and false pregnancies have been extremely common. Additonally these animals have had dry skin with signs of dermatitis (Edwards, 1975).

The main symptoms of molybdenum deficiency in poultry are reduced vigour, lowered hatchability with ginger-haired day old chicks, dermatitis especially on the hips and backs of broilers, reduced speed of feather growth, increased feather fragility, leg weakness and bone deformities (twisted beaks and lack of ocular cavity at hatching; and twisting of leg bones). In marginal deficiencies, the only signs at hatching may be a few chicks with ginger hairs around the shoulders and a few more

Table 9.4 Effect of various injections on scab severity and feathering in 1.0–1.2 kg broilers

Injection	% change in featherless area per bird in 7 days[1]	Scab sore at 7 days[2,3]
Control	− 38	0.7
CuSO4 (2 mg Cu)	+ 18	1.8
Mo (0.5 mg Mo)	−100	0.0
Cu + Mo	− 15	1.3
Na2WO4 (1 mg W)	− 18	1.0
W + Mo	− 51	0.5
Least significant differences (P = 0.05)	30	0.4

[1] At the start of each treatment, on average there were 120 cm^2 lacking in feathers per bird, on the hips and along the back.
[2] Scabs scored 0, 1, 2, 3; 0 = nil, 3 = maximum severity.
[3] Mean scab store at start, for each treatment, was 0.7 per bird

chicks exhibiting signs similar to formalin burns. In older birds a marginal deficiency will lead to narrow, straight, clear, transparent, horizontal bands across all the primary and/or all the secondary feathers (of one or both wings) from which areas the barbules have fallen. Poultry farmers in the past have seen these mild signs as evidence of chilling!

Table 9.4 shows data obtained on the prevention of the 'scabby hip' disorder of broilers through molybdenum treatment of affected birds. Thus injection of 0.5 mg Mo (as ammonium molybdate) per bird (approximately 1.0 kg in weight) completely eliminated the dermatitis scabs and resulted in complete, thick feather cover regrowing within 7 days of treatment. Copper sulphate appeared to be a far worse molybdenum antagonist than tungstate. Since feed firms have been adding 1–2.5 ppm molybdenum (sodium molybdate) to broiler diets the scabby hip and femoral bone disorder (which in some flocks was at a 20–30% incidence) has completely disappeared.

Two papers – Peterson (1974) in the USA and Bains, Mackenzie and McKenzie (1975) in Australia – had suggested that the dermatitis feathering condition associated with the 'scabby hip' syndrome is caused by the same factor which is responsible for the 'femoral head' disorder. In this latter condition, at processing of broilers, the femurs of poorly feathered scabby birds tend to break. When this occurs, the head of the femur is left in the acetabulum. The whole bone is fragile, yet soft and powdery. One can postulate, as indeed Scott, Nesheim and Young (1969) have done, that molybdenum may be involved in the prevention of other diseases such as osteoperosis and cage layer fatigue. There is considerable evidence that molybdenum is involved in the prevention of dental caries and not merely by increasing the utilisation of fluorine (Davies, 1971).

The molybdenum requirement is low, perhaps only 0.01 ppm. However there is evidence that the availability of molybdenum in feed ingredients can often be virtually nil (Leach *et al.*, 1962). Also circumstances which lower the pH of the small intestine such as a preponderance of wheat (Major, 1975) might decimate the apparent availability of any dietary molybdenum, including that added as supplementary salts.

The most important feature of the Sydney University molybdenum research is that although molybdenum deficiency resulted in severe economic losses, no effects have been observed on the normal production parameters, i.e. live weight gain, egg numbers and feed conversion efficiency, and there were only minor (non-significant!) effects on mortality.

Molybdenum Content of Feedstuffs

The molybdenum content of some animal feed ingredients is given in Table 9.5. The wide variation in content would suggest that information on batch by batch variation in molybdenum content is needed to appraise the necessity for molybdenum supplementation. Many batches of cereal contain virtually no molybdenum. Where we have obtained a measured 0.02 ppm, this is only one half of the impurities which would be added if the AR grade chemicals used in digestion contained the maximum level of molybdenum contamination permitted. The fact that molybdenum is only toxic to poultry and pigs at levels well in excess of 100 ppm (Lepore and Miller, 1965) ensures that blanket supplementations of 0.5–2.5 ppm of molybdenum are not likely to cause problems to non-ruminants However poisoning might arise if litter from poultry fed molybdenum were used on alkaline pastures fed to ruminants. Because of this, research is needed to ascertain the minimum 'blanket' supplementation effective in all instances of molybdenum inadequacies.

Table 9.5 Molybdenum content of some feedstuffs (ppm)

Barley	0.50^1; $< 0.02-1.1^3$; 0.30 ± 0.25^3
Corn	0.06^1.
Meat and bone meal	0.05^2.
Oats	1.15^1; $< 0.02-1.0^3$; 0.25 ± 0.22^3
Peanut meal	0.25^1.
Soyabean	2.5^1; 0.70^2
Wheat	1.0^1; $< 0.02-0.10^2$
Yeast (dried brewers)	1.0^1.

[1] Scott, Nesheim and Young (1969)
[2] Personal observations
[3] Todd (1972)

Trace Minerals of Unknown Significance

At Sydney University experiments have been conducted to investigate whether any of the 'minor' trace minerals reported to be essential for poultry, might be deficient in practice. To date no responses have been obtained with vanadium, chromium, nickel, stannous and cobalt supplementations of a range of poultry diets.

In one experiment a feathering response to vanadium (sodium orthovanadate) was obtained. In the same experiment an identical response was obtained to molybdate supplementation. The two effects showed no indication of being additive. Re-examining the original vanadium work of Hopkins and Mohr (1971) suggests that their synthetic diets were supplemented neither with molybdenum nor selenium compounds. As Hewitt (1974) has pointed out, the substitution of vanadium for molybdenum is varied by living organisms. For example vanadium is unable to replace molybdenum in nitrogen fixation in certain species of *Azotobacter* or in *Rhizobia* or in certain strains of *Clostridium*. However vanadium can substitute in *A. chrocococcum*, in *A. vinelandii* and in *Mycobacterium* with up to 70% efficiency. Vanadium appears to be essential in addition to molybdenum in *Scenedesmus obliquus*. If vanadium is essential for poultry, then it is the one minor trace element which would require careful monitoring. It is found mainly in the ether extract; hence levels in vegetable proteins are likely to diminish with improved oil extraction methods. It is possible that the varied response of poultry to linoleic acid sources may be due to some contaminant — such as vanadium.

Biotin Deficiency in Broilers

My own observations, together with those of Pearson *et al.* (1976) have shown that the biotin-responsive fatty liver and kidney syndrome (FLKS) in broiler chickens only results if four factors occur simultaneously.

1. Marginal biotin levels in breeder diets.
2. Breeder flocks less than 40 weeks of age.
3. Marginal biotin levels in broiler diets.
4. Some stress imposed on the broilers.

A fifth factor can probably exist — failure to synthesise biotin by intestinal bacteria — when diets predominate in wheat and animal proteins. I am convinced that failure to reproduce FLKS experimentally does not indicate any factor other than biotin deficiency being involved in FLKS but represents a failure to satisfy points 1 and 2.

Observations have been made on the biotin content of hatching eggs obtained from broiler parent stock of various ages, housed, fed, and managed identically. The broiler diets contained a basal 70 ppb biotin and were unsupplemented. The data are given in *Table 9.6*. It would appear that biotin uptake by young breeding stock is low, possibly because of the sudden elevated calcium intake around point-of-lay, reducing intestinal synthesis. The data of Hood *et al.* (1976) strongly suggest that FLKS is purely due to biotin inadequacies. Both pyruvic carboxylase and acetyl-CoA carboxylase are biotin dependent. In FLKS, only pyruvic carboxylase is deficient, due to preferential biotin utilisation by acetyl-CoA carboxylase. In biotin deficiency later in life this is not necessarily the case. The

Table 9.6 Biotin content of the yolks of broiler breeder eggs in relation to age of breeding stock. The breeding stock housed on litter were fed wheat—meat meal diets, not supplemented with biotin

Age at time of producing hatching eggs (weeks)	Mean biotin content (ppm)	Range in biotin content (ppm)
up to 30	0.34 ± 0.06	0.20 − 0.40
31 − 40	0.45 ± 0.03	0.35 − 0.55
41 − 50	0.52 ± 0.07	0.40 − 0.65
51 +	0.57 ± 0.08	0.45 − 0.70

For comparison reference may be made to Brewer and Edwards (1972), who suggested that yolk biotin contents of 0.20 ppm probably were adequate for hatching. However in looking at the mortality figures of Brewer and Edwards (1972), and the poor overall hatchability obtained, this conclusion would be very suspect.

FLKS symptoms can, I believe, be explained biochemically solely on the basis of biotin deficiency. One outbreak of FLKS occurred in Western Australia, despite biotin supplementation of the breeder and broiler diets. However, it was subsequently shown that the biotin activity in these diets was very low due to lack of incorporation of antioxidants.

Pyridoxine—Folic Acid Hatchery Syndromes

Research at Sydney University has shown that hatchery-reject chicks obtained from a clubbed down hatchery syndrome are responsive to simultaneous drenching with pyridoxine and folic acid at age 1 day. The main signs of the syndrome are lowered hatchability; reduced vigour in chicks aged 1 day, with a high proportion of the weak chicks having extremely short down on their backs (appearance like short velvet); and very short, orange-coloured, bilateral clubbed down on the abdomens, extending up to the breast (the normal chick colour is pale creamy yellow). In the worst Australian outbreak, levels of trace nutrients, especially manganese, zinc, iron, copper, B_2, B_{12}, nicotinic acid and pantothenic acid had been gradually increased in an effort to counteract the problem. Reducing the dietary levels of manganese and zinc by about 50%, and changing from their sulphates to oxides markedly alleviated this hatchery syndrome. Drenching of the day old chicks with pyridoxine dramatically improved subsequent feathering; however folic acid supplementation conjointly with the pyridoxine was necessary to lower mortality in reject chicks to acceptable levels. At the time of the problem, the breeding stock diets predominated in wheat, with meat meal as the main supplementary protein source. The breeder diets were supplemented throughout the problem with 1 ppm of folic acid and 2.5 ppm of pyridoxine. Our experiments would suggest that other trace nutrients may be involved marginally in the syndrome. At present vanadium, chromium, nickel, manganese, linoleic acid and inositol have not been completely eliminated.

A literature search failed to reveal any reported *simultaneous* deficiencies of pyridoxine and folic acid in poultry. Simple folic acid deficiencies have been recorded in broiler chickens (Saxena et al., 1954) and in turkey breeding stock (Lee, Belcher and Miller, 1965; Miller and Balloun, 1967). The main symptoms of folic acid deficiency in broiler chickens were depressed live weight, a high incidence of perosis and poor feathering. In the Australian hatchery syndrome, the reject chicks reared to 1.5—2.5 kg were peculiarly lacking in perosis (no cases in more than 2000 birds reared). None of the treatments which contained folic acid and pyridoxine (although effective in lowering mortality) had any effect on subsequent live weight gain. Although poor feathering was present in the broilers, it was responsive *only* to supplementary pyridoxine. The symptoms of folic acid deficiency in turkey flocks resulted in their poult progeny becoming nervous and suffering from 'cervical paralysis' after 1 week of age (Miller and Balloun, 1967). In Australia the broiler chickens were extremely lethargic (i.e., not nervous); however many showed signs identical with the photograph of Miller and Balloun (1967) of cervical paralysis, but in the chickens this occurred 12—72 hours after hatching.

Table 9.7 Suggested trace mineral supplementation per kg diet on wheat diets containing animal proteins

Nutrient	Probable optimal form	Broiler breeding stock	Broiler stock
Antioxidant (mg)	—	125	125
Copper (mg)	oxide	5	5
Iodine (mg)	—	1	1
Iron (mg)	—	15[1]	20[1]
Manganese (mg)	oxide	40	60
Molybdenum (mg)	—	1	1
Selenium (mg)	selenite	0.05[2]	0.05[2]
Zinc (mg)	oxide	60	40
Biotin (μg)	—	70	50[3]
Choline (mg)	—	50	200[4]
Folic acid (mg)	—	3	2[4]
Nicotinic acid (mg)	—	20	20
Pantothenic acid (mg)	—	8	5
Pyridoxine (mg)	—	10	4[4,5]
Riboflavin (mg)	—	8	4
Vitamin A (iu)	—	10000	10000[4]
Vitamin D (iu)	—	2000	2000[4]
Vitamin E (iu)	—	20	10[4]
Vitamin K (mg)	—	4	2
Vitamin B_{12} (μg)	—	6	6

[1] nil, when 10% or more of animal protein ingredients
[2] only included where selenium inadequacies (not occasional selenium-responsive conditions) are known to exist
[3] only during the broiler starter phase
[4] to be reduced by two-thirds during the finisher phase
[5] may need doubling, if high levels of copper sulphate are used for growth promotion

General Considerations of Trace Nutrient Levels

Observations at Sydney University would suggest that the dietary supplementation for those vitamins whose requirements are normally provided in part by intestinal synthesis (i.e. folic acid, pyridoxine, vitamin K, etc.) need elevating when breeding stock are being fed at levels considerably below *ad libitum* intake. However it would seem that the overall levels of trace mineral supplementation should not be raised in breeder diets. In the case of broilers growth is more rapid than when many trace nutrient requirements originally were ascertained, and feed conversion has improved markedly. If a trace nutrient requirement were calculated when the feed conversion was 4.0, now that feed conversion is around 2.0 the dietary inclusion level ought to be doubled. Our research would support this general hypothesis. However for those vitamins whose requirements are provided in part by intestinal synthesis, the overall lifetime synthesis of a broiler chicken is likely to be markedly reduced due to both the shorter growing time and to the inhibitory effects on intestinal synthesis of drugs, and even the increased dietary levels of trace minerals. *Table 9.7* shows suggested nutrient supplementation levels for broiler breeder and broiler diets bearing these points in mind.

References

Alam, S.Q. (1971). In *The Vitamins* (2nd edn.), Vol III, pp.356–394. Ed. by W.A. Sebrell and R.S. Harris
Bains, B.S. (1975). Personal communication
Bains, B.S. and Mackenzie, M.A. (1975). *Aust. vet. J.*, **51**, 364
Bains, B.S., Mackenzie, M.A. and McKenzie, R.A. (1975). *Aust. Vet. J.*, **51**, 140
Black, C.A., Evans, D.D., White, J.L., Ensminger, L.E. and Clark, F.E. (1965). *Methods of Soil Analysis*, Pt. 2 (No. 9 in series *Agronomy*). Madison, Wisconsin, USA; Am. Soc. of Agronomy
Brewer, L.E. and Edwards, H.M. (1972). *Poult. Sci.*, **51**, 619
Buck, W.B. and Evans, R.C. (1973). *Clinical Toxicology*, **6** (3), 459
Connor, J.K. (1968). 'The determination of available energy of Queensland cereal grains for poultry.' MScAgr thesis, Queensland University
Connor, J.K. and Major (1975). Personal communication
Creek, R.D., Vasaitis, V., Pollard, W.O. and Schumaier, G. (1962). *Poult. Sci.*, **41**, 901
Cruickshank, G. (1975). *Poult. World London*, **19 June**, 14
Culton, T.G. and Bird, H.R. (1940). *Poult. Sci.*, **19**, 424
Cumming (1975). Personal communication
Daghir, N.J. and Balloun, S.L. (1963). *J. Nutrition*, **79**, 279
Daghir, N.J. and Shah, M.A. (1973). *Poult. Sci.*, **52**, 1247
Davies, I.J.T. (1971). *The Clinical Significance of the Essential Biological Metals*. London; Heinemann
Davies, N.T. (1974). *Proc. Nutr. Soc.*, **33**, 293
Edwards (1975). Personal communication

Fieser, L.F. and Fieser, M. (1957). 'Introduction to Organic Chemistry.' Boston: Heath
Fisher, C., Lausen-Jones, A.P., Hill, K.J. and Hardy, W.S. (1973). *Br. Poult. Sci.*, **14**, 55
Flight, C.H. (1956). *Fmg. S. Afr.*, **32**, 37
Fuller, H.L. (1964). *Vitams. Horm.*, **22**, 659
Gardiner, M.R. (1964). *J. Agric. W. Aust.*, **5** (11), 890
Gladstones, J.S. (1971). *Dep. Agric. W. Aust. Bull.*, **3834**
Hall, S.A. (1974). *Vet. Rec.*, **94**, 42
Hartley (1975). Personal communication
Hewitt, E.J. (1974). In *Plant Biochemistry*, Vol. II of *Biochemistry Series I*, pp.201–236. Ed. by D.H. Northcote. London; Butterworths
Hood, R.L., Johnson, A.R., Fogerty, A.C. and Pearson, J.A. (1976). In *Lipids*. In press
Hopkins, L.E. and Mohr, H.E. (1971). In *Newer Trace Elements in Nutrition*. Ed. by W. Mertz and W.E. Cornatzer. New York; Marcel Dekker
Kirchgessner, M. and Friesecke, H. (1963). *Arch. Gefluegelk.*, **27**, 412
Koentjoko (1974). MScAgr thesis, Sydney University
Kratzer, F.H., Earl, L. and Chiarquanont, C. (1974). *Poult. Sci.*, **53**, 1795
Leach, R.M., Turk, D.E., Zeigler, T.R. and Norris, L.C. (1962). *Poult. Sci.*, **41**, 300
Lee, C.D., Belcher, L.V. and Miller, D.L. (1965). *Avian Dis.*, **5**, 504
Lepore, P.D. and Miller, R.L. (1965). *Proc. Soc. exp. Biol. Med.*, **118**, 115
Lillie, R.J. and Briggs, G.M. (1947). *Poult. Sci.*, **26**, 295
Lucas, H.L., Heuser, G.F. and Norris, L.C. (1946). *Poult. Sci.*, **25**, 137
Major (1975). Unpublished
McDonald, M.W., Humphris, C., Short, C.C., Smith, L. and Solvyns, A. (1969). *Proc. Aust. Poult. Sci. Conv.*, p.223
Millar, S.M. (1972). BScAgr thesis, Sydney University
Miller, D.L. and Balloun, S.L. (1967). *Poult. Sci.*, **46**, 1502
National Research Council (1971). *National Academy of Sciences Nutrient Requirements of Poultry*
Nell, J.A. (1975). Sydney Univ. Sept. 1975 Newsletter of Poult. Husb. Res. Foundation
Olumu, J.M., Robblee, A.R., Clandinin, D.R. and Hardin, R.T. (1975). *Can. J. Anim. Sci.*, **55**, 71
Packham, R.G. and Payne, C.G. (1973). *Aust. J. exp. Agric. and Anim. Husb.*, **13**, 656
Packham, R.G., Royal, A.J.E. and Payne, C.G. (1973). *Aust. J. exp. Agric. and Anim. Husb.*, **13**, 649
Payne, C.G., Gilchrist, P., Pearson, J.A. and Hemsley, L.A. (1974). *Br. Poult. Sci.*, **15**, 489
Pearson, J.A., Johnson, A.R., Hood, R.L. and Fogerty, A.C. (1976). In *Lipids*. In press
Peters, S.M. and Yule, W.I. (1974). *Proc. Aust. Poult. Sci. Conv.*, 97
Peterson, G.H. (1974). *Poult. Sci.*, **53**, 822
Queensland Dept. Prim. Ind. (1974). *Nutrient Composition of Feedstuff Ingredients*. Yeerongpilly

Robblee, A.R. and Clandinin, D.R. (1970). *Poult. Sci.*, **49**, 976
Roine, P. and Ahlström, A. (1970). *Nutritio et Dieta* No. 15, 29–37. Basle; Karger
Sarma, P.S., Snell, E.E. and Eluehjem, C.A. (1946). *J. Biol. Chem.*, **165**, 55
Sathe, B.S. (1964). PhD thesis, University of New England, Armidale, Australia
Saxena, H.C., Bearse, G.E., McLary, C.F., Blaylock, L.G. and Berg, L.R. (1954). *Poult. Sci.*, **33**, 815
Scott, M.L. (1966). *Proc. Cornell Nutr. Conf.*, 35–55
Scott, M.L., Nesheim, M.C. and Young, R.J. (1969). *Nutrition of the Chicken*. Ithaca, New York; Scott and Associates
Seelig, M.S. (1972). *Am. J. Clin. Nutr.*, **25**, 1022
Smetana, P. (1974). *Proc. Aust. Poult. Sci. Conv.*, 89–94
Smetana, P. and Morris, R.H. (1972). *Proc. Aust. Poult. Sci. Conv.*, 209–215
Todd, J.R. (1972). *J. Agric. Sci., Camb.*, **79**, 191
Turner, W. and Payne, C.G. (1971). *Aust. J. exp. Agric. and Anim. Husb.*, **11**, 629
Van Leong (1975). Personal communication
Wallis, A.S. and Parker, A.J. (1974). *Vet. Rec.*, **95**, 301
Yoshida, M. and Morimoto, H. (1960). *Vitamins*, Kyote, **21**, 324

10

FUTURE DEVELOPMENTS IN FEED COMPOUNDING IN EUROPE

C. BRENNINKMEYER
Chairman of Committee A of FEFAC

The future developments of feed compounding in Europe can be divided into two parts. Firstly there are the raw materials to be used for compounding feed for the different classes of livestock and secondly the nutritional requirements used to calculate the optimal composition of those feeds. In the future there is likely to be a shift to less common raw materials.

To feed all people of the world in an adequate way there will be a requirement for palatable foods. Cereals and even part of the vegetable proteins may well become normal ingredients of the human diet, and together with the available animal proteins and fats and table vegetables will compose the daily human food. To produce animal proteins and fats there is likely to be a move to those products which are not necessary on the human food markets. Those will be the world market surpluses of cereals and oil seeds together with the offal products not suitable for human consumption, or specially produced for animal feeds. To this last group belong tapioca and banana meal as well as fish meal and alfalfa. With regard to offal there are the meat and bone meal and offal fats which can be used in conjunction with the milling byproducts, i.e. the various types of brans, pulps and most of the oil meals. In addition to the above-mentioned agricultural and sea cultural products there is already a full scale production of synthetic products for animal feeds in the form of amino acids and single-cell proteins.

At present there seems to be ample food production for world requirements when there are no poor harvests. The problem is to transport the food to the places where it is needed and to find the money to pay for it. There is still enough scope to increase food production and keep pace with the present rate of population growth for a short period of time.

However, we must not be too optimistic and believe that in the future the rate of population growth will decrease such that the growth and production of food even in the less developed countries will keep pace with world needs. History has shown us that in the past we have had to deal with large fluctuations in the supply of raw materials. We have learned already to switch when it has been necessary. During the last years of the crises for animal production, due largely to the increase in world cereal prices, the consumption of animal products remained nearly constant. In the European Communist societies there is still an increase in the consumption of better food products. Only in the USA with its very high consumption of animal protein was there a slight decrease of consumption with very disastrous consequences for the cattle-producing sector. At the same time people in the USA have learned to produce meat at a much lower price even if they had to switch from the fat prime quality standard to a leaner one.

Some of the big fluctuations are likely to be levelled off in the future because of governmental interventions that are almost certain to take place in the next few years. Most governments have to take these measures mainly because of political reasons; they cannot afford instability of prices for the primary human needs of food, energy and housing. Another reason for less fluctuation is the support that is given by governments to the producers of agricultural products. This support will automatically lead to a maximisation of production. How difficult it is to stop this and avoid over-production we can learn from the 1 million tons excess skimmed milk powder which had to be incorporated into livestock feeds in the Common Market in 1975/76.

Being less optimistic, we shall have to look more critically at the type of animal protein we are going to produce and the methods involved. In that case we may have to concentrate on those animals which are the worst convertors of energy and protein, the ruminants. They are in fact the best, most economic producers when there is competition for food suitable for human consumption because they are able to produce independently from the human food sources. This development was stimulated during the protein crisis of 1973 and the new plants for recycling animal watse and for upgrading straw and other roughages point in that direction. Besides this a lot of feed can be saved by better management when it becomes necessary. That this is possible is proven already by the use of boars instead of male castrates, the use of hormones and so on. All such measures can produce about 20% more meat for the same amount of feed without any real problem for the consumer. *Table 10.1* gives some projected figures for the improvements that may become possible.

Another type of production that will come to attention very quickly when there is a real danger of shortage in animal protein is of course fish production. With a possible feed conversion of somewhere between 1 and 1.5:1, the fish is the best convertor available today, although the protein content of its feed has to be high and of relatively good quality. In China the solution to this problem has been achieved by cultivation of the grass carp. This fish lives on vegetables and is considered to be the ruminant of the fishes. This same fish has also recently received much attention in the USA.

The most important factor which will determine the future of the compounding industry in Europe remains, on the other hand, the consumer, since it is he who has to accept the animal products, and to pay for them. Independent of the availability of raw materials, animal productionists will have to contend with this particular influence. If they can continue to convince the consumer to eat animal products then they can look to the future. If on the contrary the big concerns which are

Table 10.1 Projected figures for food conversion efficiencies of farm livestock compared to present levels of production

	Kg compound feed per kg product		Kg 'human' protein per kg product		Kg 'human' cereal per kg product	
	Now	Future	Now	Future	Now	Future
RUMINANTS						
Milk	0.3	0–0.2	0–0.02	0	0–0.3	0
Meat	0–8	0–7	0–0.1	0	0–7	0
PIGS						
Meat	4.7	3.7	0.5	0.3	2.8	0.3
POULTRY						
Eggs	3.0	2.7	0.3	0.2	1.5	0.5
Meat	3.3	3.0	0.6	0.4	1.8	1.0
FISH						
Meat	2.0	1.5	0.6	0.5	0.2	0.1

Conversions now and in the future calculated inclusive production

producing artificial meat, milk, wine, etc. win the day, then the feed compounders can only await governmental support for plant production, and themselves turn to the production and selling of pet foods!

Trying to be more realistic about the future, there is likely to be a shift towards less normal ingredients for ruminant diets. (In 1975, in the USA alone ruminants ate more than 30 million tons of maize and nearly 4 million tons of soya which could be replaced when these feedstuffs could be more economically used elsewhere). After that there will possibly be a shift from the less good convertors, pigs and poultry towards fish. This last shift will only be needed if we cannot save more raw materials by better management and feed additives. However, both of these shifts are likely to be very gradual ones.

Parallel with the above-mentioned developments there is likely to be more economical use of the raw materials due to developments leading to a better understanding of their nutrient content. At the moment ration calculations are based on crude protein, digestible crude protein, energy, total or available calcium, phosphorus, several amino acids, crude fibre and sometimes other minerals, fatty acids and perhaps some vitamins. In the USA and in several countries of the Common Market, as well as in the Oskar Kellner Institute in Rostock, research workers are studying the value of switching to available amino acids rather than using simply crude protein or digestible crude protein for the formulation of animal rations. In the long run this change is likely to take place but at the moment it is still uncertain which are the correct figures to use. At present there is the well known *Feedstuffs* (1975) list where there is a column given for the availability of amino acids for poultry, but figures for amino acid availabilities in many feedstuffs are not yet available. The products not yet characterised are the ones with a low protein content, not normally used for poultry feeds, at least in the USA. It should therefore be possible to work with these figures for the formulation of most of the poultry feeds.

However, when one tries to do this two problems still exist. Firstly, it is very difficult to find the exact figures for requirements and secondly, the literature contains a lot of different figures for amino acid availability, some of which look more likely than those in the list. Consider, for example the availability figures for soya, fish meal, various offals and skimmed milk powder. Soya is quoted to be equal to milk protein at 98%; all the fish products come at 95%, followed by meat and bone meal at 90%. The figure for blood meal of 60% is even less than the figure for feather meal which is 65%. In addition Milo protein is 10% less available than corn (83% as against 93%). Therefore, there are several objections in accepting this list, and the figures are not yet suitable for use in practice. One of the reasons for these anomalies is the way in which the figures are collected. Whilst they are all found in the literature they are not produced by the same method nor by the same people. Most of the methods, the direct ones as well as the indirect ones, are quite satisfactory for producing a good list of figures as long as one and the same method is used in all the products tested. As soon as several methods are used, problems arise.

The studies on availability of amino acids should be quite simple if the normal methods of calculation are used to determine the amino acids in the feed and in the faeces, the so-called apparent digestibility. The true digestibility can be calculated by correcting the apparent digestibility for the excretion of metabolic or endogenous amino acids (enzymes and micro-organisms), determined with a control group on a protein-free diet. This method of correction assumes animals on the protein-rich diet excrete equal amounts of metabolic amino acids as those on a protein-free diet. In rat studies, however, Slump (1975b) found a rather constant amino acid pattern in the faeces after feeding different types of protein. Even after correction the differences were so small that he concluded that nearly all faecal amino acids originate from endogenous proteins, which themselves have a rather constant amino acid pattern. The main objection against this faecal analysis method is the possible influence of microbial activity in the caecum, colon and rectum by which unabsorbed parts of food proteins can be degraded. Some authors therefore prefer analysis of amino acids in the ileal digesta instead of in the faeces. If this is done, however, absorption of amino acids in the large intestine, if there is any, is neglected. To see which was the most promising procedure of analysis Slump (1975a) fed rats two diets, one with and one without protein, both marked with chromic acid. After one week the animals on these diets were killed. Pooled samples of intestinal contents from ileum, caecum, colon and rectum were collected and analysed. As can be seen from *Figure 10.1* there is only a slight decrease in the caecum, with the exception of cysteine, methionine, and threonine. After the caecum the decrease is quite large followed by a smaller one in the end of the large intestine.

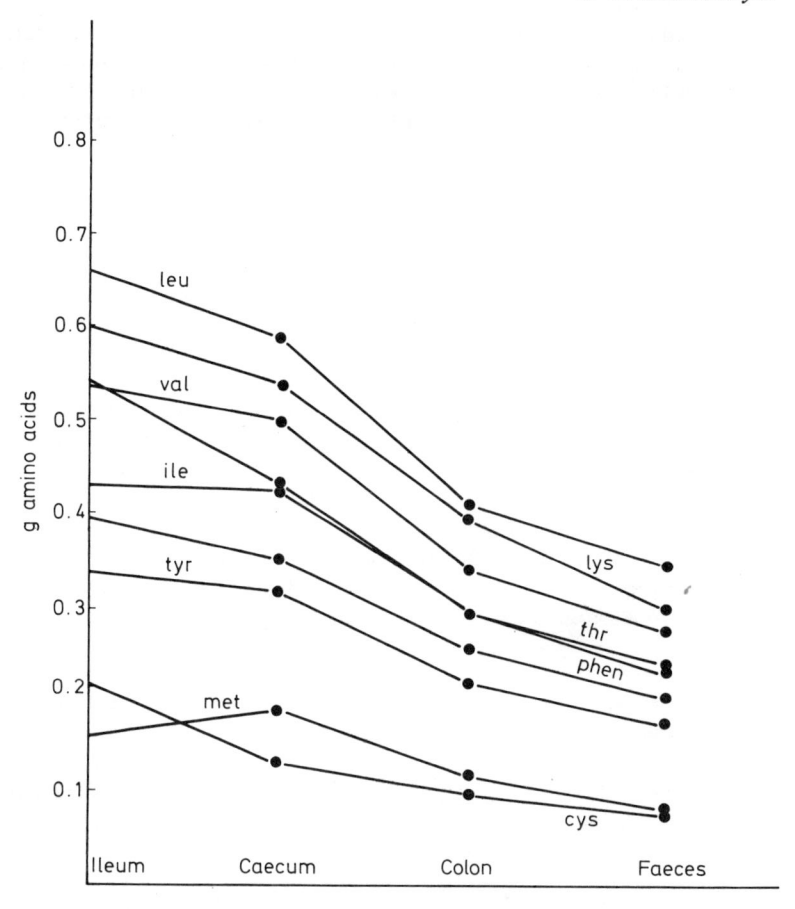

Figure 10.1 Amino acid levels related to the same amount of diet marker Cr_2O_3 in different parts of the rat intestine

In this experiment the decrease of total amino acid was greater than that of total nitrogen. This is an indication that deamination may have occurred. But because it seems illogical that this happens to the same extent for all amino acids one must conclude that there is absorption of amino acids in the large intestine. Evidence for absorption of amino acids in the large intestine was obtained by the experiments of Rettura (1972). He fed di-amino pimelic acid to rats which were deficient in lysine in their diets. Rats cannot themselves decarboxylate this acid to form lysine. The growth-promoting effect was two-thirds of that obtained when DL-lysine was added. Therefore the lysine produced by the bacteria from the di-amino pimelic acid was subsequently used by the rat. Rats which received neomycin in their drinking water to suppress bacteriological activity did not show any growth-promoting effect. When ileum analysis is used therefore the digestibility figures are too low. Whether it is correct to work with true digestibility coefficients for the amino acids depends to some extent on the kind of exogenous proteins used. If they are mainly microbial proteins, large amounts of amino acids may be synthesised in the large intestine, and this parameter is less useful. When on the other hand a major part is originating from enzymes, mucoproteins etc. excreted amino acids mean a real loss of body amino acids. In respect of the amino acid balance these lost amino acids play the same role as those from non-digested feed proteins.

In the future, a shift to available amino acids will take place in feed formulation. Even with the data now available it is possible to use these figures but the user must know the restrictions involved.

Energy systems for the future can be predicted somewhat more easily. In the last decade there have been changes in the energy content of different animal feedstuffs. Up to the present time surpluses of amino acids etc. have been added but when there is a better knowledge about the exact requirements for the animal together with an exact level in feeds produced it will become very important to reassess the exact amount of energy in that feed to get the correct energy-to-protein ratio.

Many systems now used in the diverse sections of the industry are much better than the old ones for energy evaluation, but they are not ideal for all circumstances. The systems will be considered briefly. The main systems are given in *Table 10.2*.

Table 10.2 Main energy systems used in animal nutrition

	Ruminants	*Pigs*	*Poultry*
NRC	NE gain, maintenance, lactation TDN	ME (DE)	ME
Rostock	NEF_R	NEF_S	NEF_P
ARC	ME (Kla, Km, Kf) SE	DE	ME

NRC

Many people still work with TDN in the United States, but there is a case for a switch to NE for growth and maintenance in beef cattle and to NE for lactation and maintenance for dairy cattle.

DDR (ROSTOCK)

As a logical follow-up to the work from Oskar Kellner the DDR people developed their system of net energy fat for the different groups of livestock. They assume that in normal rations there is not much difference in the relative utilisation of the digestible nutrients for the various purposes of growth, maintenance and lactation compared with fat formation. A sign that they are thinking of a separate value for lactation is given by Hoffman (1972). He published a regression equation for dairy cows with a higher coefficient for protein and carbohydrates and a lower one for fats and fibre.

ARC

This system is based on the ME content of the feedstuffs, the values being additive. Depending on the concentration of the feed and on the form of production the total ME is converted into NE by special factors.

For pigs things are much simpler with the NE and DE and the ME systems as much as possible based on digestibility figures for pigs. Poultry is the only selection that is uniform. Almost everyone is using the ME systems based on trials where this ME is directly determined by bomb calorimetry.

All of these systems try to predict as accurately as possible the relationship between feed and the production realised with it. In Belgium, Buysse (1974) did comparative studies of the several systems for dairy cattle and for fattening beef cattle. The variation was from −30% to +10% for the different systems for fattening cattle with the corrected ADAS system (empty bodyweight and meaty type) being the best with a value of 1.2%. For dairy cattle all systems overestimated the production, with the NE lactation system using the coefficients of van Es (1974) giving the best values at 1.4% and the Rostock system the worst with 21.8%.

Energy figures for the various classes of animal and types of production will be produced. It is in most cases only another column for the computer input. For prediction of the production it is easier to work with the NE system but one can perhaps use the ME system, then translate into NE.

The future will also bring some more up-to-date figures for vitamins. At the present time everyone is using ample amounts of most of the vitamins compared with the actual requirements. The Common Market regulations, which make the declaration of some of the vitamins obligatory, have stimulated the overdosing very much because it looks nice and the costs are still relatively low. In the future these additions will become more realistic. At the same time there will be more known about requirements for some of the old vitamins. Also new forms of vitamins, for instance the vitamin D25 hydroxy analogue, will come onto the market.

The same applies for the minerals and the trace elements. The high cost of phosphorus today has already taught feed manufacturers to be careful about overinclusion of this mineral. On the other hand practice has shown the disastrous financial consequences of underdosing. The method of keeping the animals has a very big influence on the required level of minerals and trace elements in the feed. A lot of studies going on at present on the availability of phytate phosphorus may demonstrate exactly what changes are likely in the future.

Recent evidence is demonstrating the importance of sodium, potassium and magnesium. Very little calculation of inclusion rates is carried out at the moment but this is bound to change in future years. The number of trace elements will also be extended in the coming years. Selenium has already been included in many animal feeds; molybdenum may follow, and perhaps some more, like borium, may be added, as the composition of animal feeds and types of feeding systems change.

The last group not yet mentioned comprises the fatty acids. Linoleic acid has strong competition as it is suitable for human feeds. In the future with fewer or no cereals at all included in animal rations, linoleic acid will be needed, but the price will be prohibitive against any overdosing so that the requirements of this type of material will have to be determined accurately.

References

Feedstuffs (1975). Volume 47, No. 38, p.33
Slump, P. (1975a). In *Protein Nutrition. Quality of Foods and Feeds. Part I.* Ed. by M. Friedman. New York; M. Dekker
Slump, P. (1975b). Unpublished
Rettura, G. (1972). IX Congrès international de Nutrition, Mexico
van Es, A. (1974). In press
Buysse, F.X. (1974). R.V.V. Nieuwere inzichten omtrent de Energetische voeder evaluatie en daarmede aansluitende voedernormen 5.

PARTICIPANTS

The tenth Nutrition Conference was organised by the following committee:

Dr. R.J. Andrews (Rank, Hovis and MacDougall Ltd.)
Dr. L.G. Chubb (Spillers Ltd.)
Mr. N.H. Cuthbert (BOCM Silcock Ltd.)
Mr. D.G. Filmer (Dalgety Crosfields Ltd.)
Dr. J.T. Morgan
Mr. J.R. Pickford (BP Nutrition (UK) Ltd.)
Mr. G.B. Plowman (G.W. Plowman & Son Ltd.)
Mr. P. Pratt (Eastern Counties Farmers Ltd.)
Mr. R.E. Pye (W. & J. Pye Ltd.)

Dr. K.N. Boorman
Dr. D.J.A. Cole
Dr. W. Haresign (Secretary)   } University of Nottingham
Professor G.E. Lamming
Professor D. Lewis (Chairman)
Dr. H. Swan

The tenth Conference was held at the School of Agriculture, Sutton Bonington, from 4th to 6th January, 1976 and the committee would like to thank the various authors for their valuable contributions.

The University of Nottingham is grateful to BP Nutrition (UK) Ltd. for the support that they have given in the organisation of this Conference.

The following persons registered for the meeting:

Alderman, Mr. G.	ADAS, SE Region, Block A, Coley Park, Reading RG1 6DT
Alford, Mr. J.	Harper Adams College, Newport, Salop.
Alston, Mr. J.	Scottish Agricultural Industries Ltd., 25 Ravelston Terrace, Edinburgh, EH4 3ET
Andrews, Dr. R.J.	RHM Agriculture (Exports) Ltd., Deans Grove House, Deans Grove, Colehill, Wimborne, Dorset BH21 7AE
Angastiniotis, Mr. E.K.	Spillers International Agriculture Ltd., 4–6 Cannon Street, London EC4M 6EB

Participants

Arnott, Mr. J.	Dalgety Crosfields Ltd., Crosfields House, The Promenade, Clifton, Bristol BS8 3NJ
Banton, Mr. C.	Trouw (Great Britain) Ltd., The Mill, Harston, Cambridge CB2 5NL
Barber, Mr. R.A.	Kew House Farm Ltd., 142 Southport Road, Scarisbrick, Southport
Barber, Mr. W.P.	ADAS, MAFF, Coley Park, Reading
Batty, Miss R.I.	J. Bibby Agriculture Ltd., Richmond House, Rumford Place, Liverpool L3 9QQ
Beaumont, Mr. D.	Cooper Nutrition Products Ltd., Stepfield, Witham, Essex
Beatty, Mr. D.W.J.	ICI Agricultural Division, Protein Department, Billingham, Cleveland
Beer, Mr. J.H.	W.P. Monkhouse & Sons Ltd., Low Mill, Langwathby, Penrith, Cumbria
Berry, Mr. M.H.	Cooper Nutrition Products Ltd., Stepfield, Witham, Essex
Bindloss, Mr. A.	Ashridge, Middletown Lane, East Budleigh, Devon
Bishop, Mr. A.	Morning Foods Ltd., North Western Mills, Crewe, Cheshire
Blaxter, Dr. K.L.	Rowett Research Institute, Bucksburn, Aberdeen
Bouchard, Mr. K.A.	Cooper Nutrition Products Ltd., Stepfield, Witham, Essex
Brenninkmeyer, Dr. C.	Hendrix Voeders BV, Veerstraat 38, Boxmeer, Netherlands
Brooks, Dr. P.H.	Seale-Hayne College, Newton Abbot, Devon TQ12 6NQ
Broster, Dr. W.H.	NIRD, Shinfield, Reading, Berks. RG2 9AT
Brown, Mr. F.S.D.	Frank Wright (Feed Supplements) Ltd., Ashbourne, Derbys.

Participants

Bruce, Mr. K.F.	Wyatt & Bruce Ltd., The Mills, Bovey Tracey, Devon
Bryden, Mr. A.L.	Wm. Grant & Sons Ltd., Distillers, Girvan, Ayrshire
Buckley, Mr. K.	Pedigree Petfoods, Melton Mowbray, Leics.
Burr, Mr. A.C.	Pauls & Whites Foods Ltd., Eagle Mill, Helena Road, Ipswich, Suffolk
Burt, Dr. A.W.A.	Burt Research Ltd., 23 Stow Road, Kimbolton, Huntingdon PE18 0HU
Burt, Mr. R.J.	British Sugar Corporation Ltd., Central Offices, PO Box 26, Oundle Road, Peterborough PE2 9QU
Bush, Mr. T.J.	Volac Ltd., Crayden Old Farm, Wendy, Royston, Herts.
Buttery, Dr. P.J.	University of Nottingham School of Agriculture
Campbell, Mr. C.	US Feed Grains Council, 28 Mount Street, London W1Y 5RB
Care, Prof. A.D.	Dept. of Animal Physiology and Nutrition, Kirkstall Laboratories, Vicarage Terrace, Leeds LS5 3HL
Cassidy, Mr. J.C.	Cooper Nutrition Products Ltd., Stepfield, Witham, Essex
Chaplin, Mr. R.W.	Colborn Vitafeeds Ltd., Sheepy Mill, Atherstone, Warwickshire
Charles, Dr. D.R.	MAFF, ADAS, Shardlow Hall, Shardlow, Derby DE7 2GN
Clark, Mr. R.D.	Beecham Animal Health, Beecham House, Brentford, Middx.
Clark-Monks, Mr. R.	Nitrovit Ltd., Nitrovit House, Dalton, Thirsk, N. Yorks
Close, Dr. W.H.	ARC Institute of Animal Physiology, Babraham, Cambridge CB2 4AT

Participants

Cooke, Dr. B.	Dalgety Crosfields Ltd., Crosfields House, The Promenade, Clifton, Bristol BS8 3NJ
Coomb, Mr. A.G.	Cyanamid of Great Britain Ltd., Fareham Road, Gosport, Hants.
Corse, Dr. D.A.	Cooper Nutrition Products Ltd., Stepfield, Witham, Essex
Cox, Mr. S.	Farmers Weekly, Surrey House, 1 Throwley Way, Sutton, Surrey
Crawford, Mr. T.H.	James Clow & Co Ltd., Prince's Dock, Mills, Belfast BT1 3AD
Cruickshank, Mr. I.	Spencers Feed Supplements Ltd., Farburn Industrial Estate, Dyce, Aberdeen
Curran, Dr. M.K.	Wye College, Ashford, Kent
Cuthbert, Mr. N.H.	BOCM Silcock, Basingview, Basingstoke, Hants.
Darashah, Mr. P.N.	TRACE, Trace House, Cromwell Park, Over Cambridge CB4 5PX
Davies, Dr. J.L.	Colborn Vitafeeds Ltd., Sheepy Mill, Atherstone, Warwickshire
Dawkins, Mr. C.W.C.	MAFF, Staplake Mount, Starcross, Exeter EX6 8PE
Dawson, Dr. R.	Dalgety Crosfields Ltd., Crosfields House, The Promenade, Clifton, Bristol BS8 3NJ
Deverell, Mr. P.	BASF, UK Ltd., PO Box 4, Earl Road Cheadle Hume, Cheshire SK8 6QG
Dixon, Mr. D.H.	Brown & Gillmer Ltd., Seville Place, Dublin 1
Donnison, Mr. K.G.	Smith Klime & French, Welwyn Garden City, Herts.
Drysdale, Dr. A.D.	BP Chemicals International Ltd., Greenfield House, 69/73 Manor Road, Wallington, Surrey

Dunton, Mr. R.	Spencers Feed Supplements Ltd., Farburn Industrial Estate, Dyce, Aberdeen
Earle, Mr. G.	Stockvet Ltd., Station Mills, Nafferton, Driffield
Eddie, Dr. S.M.	BOCM Silcock, Basingview, Basingstoke, Hants.
Edwards, Mr. G.	Procter & Gamble, Ind. Chemical Sales Department, PO Box 9, Hayes, Middx.
Elliott, Mrs. K.M.	Pauls and Whites Food Ltd., Lords Meadow Mill, Crediton, Devon
Ellis, Dr. N.	SAI Ltd., 25 Ravelston Terrace, Edinburgh EH4 3ET
Emmans, Mr. G.C.	East of Scotland College of Agriculture, Kings Buildings, West Mains Road, Edinburgh
Ennis, Mr. M.J.	Mixrite (I) Ltd., Bennettsbridge, Co. Kilkenny, Ireland
Evans, Dr. P.J.	Unilever Research Ltd., Colworth House, Sharnbrook, Bedford
Fairbairn, Dr. C.B.	MAFF, Block C, Brooklands Avenue, Cambridge
Fairley, Mrs. C.	J. Bibby Agriculture Ltd., Richmond House, 1 Rumford Place, Liverpool L3 9QQ
Filmer, Mr. D.G.	Dalgety Crosfields Ltd., Crosfields House, The Promenade, Clifton, Bristol BS8 3NJ
Fordyce, Mr. J.	RHM Animal Feed Services Ltd., Deans Grove House, Colehill, Wimbourne, Dorset
Foye, Mr. S.H.C.	AITS, 156 Oxford Road, Reading
Grant, Mr. J.I.	Farmway Ltd., Cock Lane, Piercebridge, Nr. Darlington, Co. Durham
Griffiths, Mr. R.J.	Pauls & Whites Foods Ltd., Radstock, Bath

188 *Participants*

Hall, Mr. G.R.	RHM Animal Feed Services Ltd., Deans Grove House, Colehill, Wimborne, Dorset
Hardy, Dr. B.	Cooper Nutrition Products Ltd., Stepfield, Witham, Essex
Haresign, Mr. W.	University of Nottingham School of Agriculture
Harker, Mr. K.	Stockvet Ltd., Nafferton, Driffield
Harvey, Mr. J.A.	Elanco Products Ltd., Broadway House, The Broadway, Wimbledon, London SW19 1RR
Heard, Mr. T.W.	The Hale Veterinary Group, 19 Langley Road, Chippenham, Wilts. SN15 1BS
Hemingway, Prof. R.G.	Glasgow University Veterinary School, Bearsden, Glasgow
Hesketh, Mr. H.R.	Sun Valley Feed Mill, Tram Inn Ind. Estate, Allensmore, Hereford HR2 9AW
Hill, Mr. B.E.	Midland Shires Farmers Ltd., Defford Mill, Earls Croome, Nr. Worcester
Hill, Mr. M.E.L.	Cooper Nutrition Products Ltd., Stepfield, Witham, Essex
Hoey, Mr. C.C.	Wyatt & Bruce Ltd., The Mills, Bovey Tracey, Devon
Hollows, Mr. I.W.	Cooper Nutrition Products Ltd., Stepfield, Witham, Essex
Holme, Dr. D.W.	Pedigree Petfoods, Mill Street, Melton Mowbray, Leics.
Holmes, Dr. C.W.	Massey University, Palmerston North, New Zealand
Holmes, Mr. J.J.	Research & Development, E.B. Bradshaw & Sons Ltd., Bell Mills, Driffield, North Humberside YO25 7XL
Hosken, Mr. E.E.	MAFF, ADAS, Block B, Government Buildings, Brooklands Avenue, Cambridge CB2 2DR

Hopwood, Mr. J.B.	Welsh Agricultural College, Llanbadarn Fawr, Aberystwyth, Dyfed
Howard, Mr. A.J.	Procter & Gamble Ltd., Whitley Road, Longbenton, Newcastle upon Tyne NE12 9TS
Howie, Mr. A.D.	Midland Shires Farmers, Defford Mill, Earls Croome, Worcester
Hudson, Mr. K.A.	Beecham Animal Health, Beecham House, Brentford, Middx.
Hunt, Mr. M.	Feed Service (Livestock) Ltd., Hartham, Corsham, Wilts. SN13 0QB
Hutchinson, Mr. H.E.	Marfleet Refining Co. Ltd., Hedon Road, Hull
Hutton, Dr. K.	Colborn Ani-Med Ltd., Heanor Gate, Heanor, Derbys.
Jackson, Dr. P.	Beecham Animal Health, Beecham House, Brentford, Middx.
Jameson, Mr. D.W.	Kemin Europa (UK) Ltd., 40–42 High Street, Maidenhead, Berks.
Jardine, Mr. G.	Aynsome Laboratories Ltd., Kentsbank, Grange over Sands, Cumbria
Johnson, Mr. J.S.	South Shropshire Farmers Ltd., Farmore Mills, Craven Arms, Shropshire
Jones, Dr. A.S.	Rowett Research Institute, Bucksburn, Aberdeen
Kennedy, Mr. G.	BASF UK Ltd., PO Box 4, Earl Road Cheadle Hume, Cheshire SK8 6QG
Kenyon, Mr. P.	Cooper Nutrition Products Ltd., Stepfield, Witham, Essex
Keys, Mr. J.	J.E. Hemmings & Son Ltd., Barford Mills, Nr. Warwick
Kidd, Mr. A.G.	ICI Agricultural Division, Protein Department, Billingham, Cleveland

Participants

Kitchen, Dr. D.I.	Pauls & Whites Foods Ltd., Research & Advisory Dept., New Cut West, Ipswich, Suffolk, IP2 8HP
Knight, Mr. D.	DAIS Ltd., 51 Salem Street, Shirley, Southampton
Kubasek, Mr. F.O.T.	ICI Ltd., Jealott's Hill Research Station, Bracknell RG12 6EY
Laws, Dr. B.M.	Pauls & Whites Foods Ltd., Research & Advisory Department, New Cut West, Ipswich, Suffolk IP2 8HP
Lawson, Mr. J.A.	Moy Park Ltd., Main Street, Donaghmore Co. Tyrone
Lea, Mr. J.E.	Morning Foods Ltd., North Western Mills, Crewe, Cheshire
Lewis, Prof. D.	University of Nottingham School of Agriculture
Lewis, Dr. M.	The Edinburgh School of Agriculture, West Mains Road, Edinburgh EH9 3JG
Lindeman, Mr. M.A.	BOCM Silcock Ltd., Basingview, Basingstoke, Hants.
Loane, Dr. D.J.	BOCM Silcock Ltd., Basingview, Basingstoke, Hants.
Lowe, Dr. R.A.	Colborn Group Ltd., Barton Mills, Canterbury, Kent
McCallum, Mr. J.M.	Cooper Nutrition Products (NI) Ltd., Sussex Street, Belfast BT15 1GH
McKendry, Mr. J.	John Thompson & Sons Ltd., Donegall Quay Mills, Gamble Street, Belfast BT1 3AK
McLean, Mr. D.R.	R.J. Seaman & Sons Ltd., Elmham Mills Dereham, Norfolk
Marangos, Dr. A.	Spencers Feed Supplements Ltd., Farburn Industrial Estate, Dyce, Aberdeen
Merk, Dr. W.	Degussa Wolfgang, GB Chemie/Anwendungstechnik, D-6450 Hanau, Postfach 602

Participants

Midgley, Mr. M.	Format, Chobham Road, Sunningdale, Berks.
Mitchell, Dr. R.M.	Spencers Feed Supplements Ltd., Wellheads Road, Farburn Industrial Estate, Dyce, Aberdeen AB2 0HG
Modderman, Dr. J.G.	Windmill Holland BV, PO Box 58, Vlaardingen, Netherlands
Moore, Mr. D.R.	Croda Agricultural Ltd., Barbers Road, Stratford, London E15 2PH
Morgan, Mr. D.	
Morgan, Dr. J.T.	The Chestnuts, Evercleech, Shepton Mallet, Somerset BA4 6DU
Mount, Prof. L.E.	ARC Institute of Animal Physiology, Babraham, Cambridge
Mudd, Dr. A.J.	ICI Ltd., Agricultural Division, PO Box 1, Billingham, Cleveland TS23 1LB
Murray, Mr. J.	Peter Hand (GB) Ltd., Russell House, 59 High Street, Rickmansworth, Herts.
Nott, Dr. H.	Nickerson Group, Rothwell, Lincoln
Oldham, Dr. J.D.	NIRD, Shinfield, Reading RG2 9AT
O'Neill, Dr. B.	Devenish Feed Supplements, 61–75 Corporation Street, Belfast
Owers, Dr. M.J.	Pauls & Whites Foods Ltd., Research & Advisory Dept, New Cut West, Ipswich, Suffolk IP2 8HP
Parker, Mr. C.G.S.	Pauls & Whites Foods Ltd., Research & Advisory Dept, New Cut West, Ipswich, Suffolk IP2 8HP
Patterson, Mr. D.	Barkers & Lee Smith Ltd., Barkers Mill, Lincoln
Payne, Prof. C.G.	University of Sydney, Camden, New South Wales 2570, Australia
Perry, Mr. F.G.	Trouw (GB) Ltd., The Mill, Harston, Cambs.

Participants

Phillips, Mr. G.	W.J. Oldacre Ltd., Technical Division Bishop's Cleeve, Glos.
Pickard, Dr. D.W.	University of Leeds, Dept. of Animal Physiology, Kirkstall Laboratories, Vicarage Terrace, Leeds
Pickford, Mr. J.R.	Cooper Nutrition Products Ltd., Stepfield, Witham, Essex
Pike, Dr. I.H.	IAFMM, Hoval House, Mutton Lane, Potters Bar, Herts. EN3 3AR
Pilbrow, Dr. P.J.	Pilwood Feeds Ltd., East Wellow, Nr. Romsey, Hants. SO5 0ZU
Pinson, Mr. D.J.	B.G. Wyatt Ltd., Chard, Somerset
Plowman, Mr. G.B.	G.W. Plowman & Son Ltd., South Holland Mills, Spalding, Lincs. PE11 1TP
Portsmouth, Mr. J.	Peter Hand (GB) Ltd., Russell House, 59 High Street, Rickmansworth WD3 1EZ, Her
Pratt, Mr. P.D.	Eastern Counties Farmers Ltd., 86 Princes Street, Ipswich, Suffolk
Putnam, Mr. M.E.	Roche Products Ltd., 318 High Street North, Dunstable, Beds. LU6 1BG
Pye, Mr. R.E.	W. & J. Pye Ltd., Fleet Square, Lancaster LA1 1HA
Randall, Miss E.M.	ADAS, Boxworth EHF, Boxworth, Cambridge
Reed, Mr. S.A.	Marfleet Refining Co. Ltd., Hedon Road, Hull HU9 5NJ, N. Humberside
Reeve, Mr. J.	Preston Farmers Ltd., County Mills Ruskington, Sleaford, Lincs.
Richardson, Mr. W.	Rose Neath, Ramsey Road, Laxey, Isle of Man
Rigg, Mr. G.	W. & J. Pye Ltd., Fleet Square, Lancaster LA1 1HA
Robb, Dr. J.	Unilever Research Laboratory, Colworth House, Sharnbrook, Beds.

Roberts, Mr. P.T.	Pauls & Whites Foods Ltd., Industrial Estate, Winsford, Cheshire
Robertson, Mr. W.S.	Cooper Nutrition Products (NI) Ltd., Sussex Street, Belfast BT15 1GH
Rosen, Dr. G.	17A Colinette Road, London SW15
Round, Mr. J.S.K.	Nitrovit Ltd., Nitrovit House, Dalton Thirsk, N. Yorks.
Rutledge, Mr. W.A.	BOCM Silcock (NI), 35/39 York Road, Belfast
Scott, Mr. L.J.	Colborn International, Barton Mills Canterbury, Kent
Scott, Mr. R.T.M.	W. & C. Scott Ltd., Excelsior Mills, Omagh, Co. Tyrone BT79 7AG
Shillam, Dr. K.W.G.	Huntingdon Research Centre, Huntingdon, Cambs. PE18 6ES
Silcock, Mr. R.	Dalgety Franklin Ltd., Dalgety House, High Street, Biggleswade, Beds.
Smith, Mr. D.	Spillers Farm Feeds Ltd., St. Giles House, 180 High Holborn, London WC1V 7AB
Smith, Mr. G.H.	Pauls & Whites Foods Ltd., Research & Advisory Dept, New Cut West, Ipswich, Suffolk IP2 8HP
Smith, Mr. T.L.	Cooper Nutrition Products Ltd., Stepfield, Witham, Essex
Speight, Mr. D.	Nitrovit Ltd., Nitrovit House, Dalton, Thirsk, N. Yorks. YO7 3JE
Stainsby, Mr. A.K.	Brandsby Agricultural Trading Association Ltd., Railway Street, Malton, N. Yorks.
Statham, Mr. R.	Criddle & Co., J & T. Peters, Glazebury, Warrington
Swan, Dr. H.	University of Nottingham School of Agriculture

Participants

Sykes, Dr. A.	Wye College (University of London), Ashford, Kent
Taylor, Dr. A.J.	Unilever Ltd., Colworth House, Sharnbrook, Bedford MK44 1LQ
Thickett, Mr. W.S.	Barhill Farm, Tushingham, Whitchurch, Salop.
Thomas, Mr. G.	Favor Parker Ltd., Stoke Ferry, Norfolk
Thompson, Mr. D.	Ranks (Ireland) Ltd., Dock Road, Limerick, Ireland
Thompson, Mr. D.E.M.	Cooper Nutrition Products Ltd., Stepfield, Witham, Essex
Thompson, Dr. F.	Rumenco Ltd., Stretton House, Derby Road, Burton on Trent DE13 0DW
Thompson, Mr. R.J.	Preston Farmers Ltd., Kinross New Hall Lane, Preston PR3 0AE
Thornton, Mr. D.D.	SADC, Bush Estate, Penicuik, Midlothian EH26 0PZ
Tonks, Mr. W.P.	Golden Vale Food Products, 124 Finchley Road, London NW3 5HT
Trapnell, Dr. M.G.	Dalgety Crosfields Ltd., Crosfields House, The Promenade, Clifton, Bristol BS8 3NJ
Treacher, Dr. R.J.	Dept. Functional Pathology, IRAD Compton, Newbury, Berks.
Waddoup, Mr. P.L.	S. Cats Feed Mill, Michel Dever Station, Hants.
Walker, Dr. T.	BP Proteins Ltd., Britannic House, Moor Lane, London EC2Y 9BU
Warren, Mr. A.J.	Barkers & Lee Smith Ltd., Barkers Mills, Lincoln
Waterworth, Mr. D.G.	ICI Ltd., Jealotts Hill Research Station, Bracknell RG12 6EY
Watt, Mr. J.A.	W. & J. Pye Ltd., Fleet Square, Lancaster LA1 1HA

Weeks, Mr. R.H.	Pauls & Whites Foods Ltd., Gregson Lane, Hoghton, Preston, Lancs. PR5 0DN
Widdowson, Mrs. V.	17A Colinette Road, London SW15
Wilby, Mr. D.T.	W.F. Tuck & Sons Ltd., The Mills Burston, Diss, Norfolk
Wilcox, Mr. N.J.	Pauls & Whites Foods Ltd., 141 Wincolmlee, Hull
Wilkins, Mr. D.A.	Seemeel Ltd., Wolsley Works, Stratford, London E15 2DX
Williams, Mr. D.J.	James Duke & Son Ltd., Abbey Mill, Bishops Waltham, Southampton SO3 1GN
Wilson, Prof. P.N.	BOCM Silcock Ltd., Basingview, Basingstoke, Hants.
Witt, Mr. G.T.	Seemeel Ltd., Wolsley Works, Carpenters Road, London E15 2DX
Wollaston, Mr. J.G.	T. Marsden & Sons Ltd., Globe Mill, Midge Hall, Leyland, Preston PR5 3TP
Wyldes, Mr. N.	BASF (UK) Ltd., PO Box 4, Earl Road, Cheadle Hulme, Cheshire SK8 6OG

Post Graduate Students

Kaushal, Mr. J.R.	University of Nottingham School of Agriculture
Hawkes, Miss A.	University of Nottingham School of Agriculture
Mortimore, Miss J.	University of Nottingham School of Agriculture

INDEX

Ambient temperature, 17
  energy intake, 18
  protein requirements, 25
  vitamin requirements, 26
  water requirements, 27
Amino acids, 103, 109, 156, 177
  absorption, 179
  availability, 178
  digestibility, 178
Apparent digestible organic matter (ADOM), 133, 135

Biotin, 169
  deficiency syndromes, 169
  in eggs, 170
Bronchitis, 159

Caesin, 139
Calcium, 106, 109, 113
  absorption, 115
  adaptation, 114
  availability, 116
  blood levels, 117, 118, 119
  requirement, 117
Carcase composition, 68
Carcase quality, 90
Climate, 1
  direct effects of, 1, 2
  indirect effects of, 1, 2
Cold, adaptation to, 14
Conduction, heat loss by, 4, 52
Convection, heat loss by, 4, 52
Coprophagic pellets, composition of, 96
Coprophagy, 95

Cottonseed meal,
  composition of, 156
  feeding syndromes, 158
Critical temperature, 11–13, 22
  lower, 4, 53, 55, 63–66
  upper, 54, 55, 65–66

Daily gain, 80
Digestible energy (DE), 75, 76, 77, 78, 180
  concentration, 79
  intake, 79

Egg production, 26, 40
  economics of, 40
  effect of protein intake, 38
Egg weight, 40
Energy,
  bodyweight gain, 35
  conservation, 84
  content of eggs, 35
  conversion efficiency, 84
  intake and milk production, 130
  metabolism, 67
  partition of, 81
  requirements, 18, 98, 99, 100
  systems, 75–78, 180–182
Energy/protein ratio, 89
Evaporation, heat loss by, 5, 52–53
External insulation, 9–12

Faeces, composition of, 96
Fat,
  deposition, 84
  requirements, 98, 109

Index

Fatty acids, 182
Feather loss score, 45–46
Feed additives, 108
Feeding regime, 34
  egg production, 41–43
  egg weight, 43
  feed intake, 43
Fibre requirements, 97, 109
Fish production, 176
Fleece insulation, 7, 9
Folic acid, 159
  deficiency syndromes, 170
Food conversion ratio, 81, 176
Food intake,
  prediction of, 45
  effect of temperature, 6, 40, 68
Food requirements, 12

Gluconeogenesis, 139
Glucose requirements, 139
Growth rate, 69, 100

Hair coat, 12
Heat deficit, 85
Heat exchange, 51
Heat loss, 5, 35, 52
  bedding, 60, 85
  calculation of, 57
  fatness, 60, 62
  floor type, 60, 85
  group size, 60, 62
  invariant, 4
  live weight, 62
  rain, 10
  sensible, 4, 53
  wind, 60
Heat production, 5, 52, 56
  feeding level, 3, 56
  temperature, 3, 59
  units of measure, 2, 3
Hypocalcaemia, 113, 119–121
Hypomagnesaemia, 120

Insulation, 5
  tissue, 6
  external, 6, 9
Iron, requirements for, 106, 109

Lupinine poisoning, 158
Lupinosis, 158
Lupinseed,
  composition, 156
  deficiency syndromes, 158

Maintenance,
  energy, 35, 81, 99
  protein, 102
Meat and bone meal,
  composition, 156
  deficiency syndromes, 159
Metabolisable energy (ME), 75–78, 156, 181
  temperature and intake of, 18, 33–37, 63–66
Milk,
  composition, 130
  fever, 113, 119–121
  replacers, 105
Molybdenum,
  contents in feedstuffs, 168
  deficiency syndromes, 166
  supplementation, 168

Net energy (NE), 75, 76, 180

Offal, 175
Organic matter digestibility, 135

Parathyroid hormone (PTH), 114, 118
Phosphorus, 119–121
  milk fever, 118–121

Phosphorus (Cont.)
  requirements for, 106, 109
  vitamin D, 117
Protein,
  absorption, 137
  degradation, 136
  deposition, 82, 83
  digestibility, 125
  intake, 83, 145
  metabolism, 132, 143–145
  milk yield, 127
  requirements, 25, 89, 101, 124, 129–149
  supply, 136–138
  temperature effects, 68
  turnover, 82
  utilisation, 138
Protein–energy interrelationships, 125
Protein–energy ratio, 128
Pyridoxine, 159
  deficiency syndromes, 170
  folic acid, 170
  supplementation, 171

Rabbits, 93
  digestive tract, 94
  growth, 100, 103
  industry, 93
  lactation, 100, 102
  reproduction, 100, 102
Radiation, heat loss by, 4, 7, 53, 61
Rain, effects on insulation, 9–12
Rapeseed,
  composition, 156
  deficiency syndromes, 160
Rectal temperature, 5, 46
Rice bran,
  composition, 156
  deficiency syndromes, 161
Rumen,
  bacterial growth, 133
  energy requirements, 133
  nitrogen synthesis, 133
  nitrogen yield, 134

Salt, requirements, 106, 109
Skimmed milk powder, 176

TDN, 77–78, 99, 180
Temperature,
  adaptation, 22
  carcase composition, 68
  carcase quality, 90
  energy metabolism, 67
  food intake, 68
  growth, 6, 69, 87
  lower critical, 4, 53, 63–66, 86–87
  mineral requirements, 70
  protein metabolism, 68
  vitamin requirements, 70
  water requirements, 70
Thermoneutral zone, 21, 53
Tissue energy loss, 23
Tissue insulation, 8
Trace mineral supplementation, 171
Trace nutrient,
  interactions, 165–166
  levels, 162
  requirements, 162, 164–165
True digestible organic matter (TDOM), 133, 135

Urea feeding, 145

Ventilation rate, 33, 38
  house temperature, 47
  performance, 38
Vitamin requirements, 26, 70, 107, 109, 171

Water loss, 27–28
Water requirements, 27, 70, 107
Wheat,
  composition, 156
  deficiency syndromes, 161–162

Wheat (*Cont.*)
  trace mineral supplementation, 171
Wind,
  critical temperature, 65–66
  heat loss, 60
  insulation, 9–10